CYBER RISK LEADERS

GLOBAL C-SUITE INSIGHTS

LEADERSHIP & INFLUENCE IN THE CYBER AGE

BY
SHAMANE TAN

First published in 2019 by MySecurity Media Pty Limited
www.mysecuritymedia.com

Cyber Risk Leaders
www.cyberriskleaders.com

National Library of Australia Cataloguing-in-Publication entry

Author	Tan, Shamane	
Title	Cyber Risk Leaders: Global Global C-suite Insights, Leadership & Influence In The Cyber Age	
ISBN	978-0-6485953-1-1	Print
ISBN	978-0-6485953-6-6	eBook

Executive Production	Chris Cubbage and David Matrai
Editing	Ann Longmore-Etheridge
Cover design	Melissa ten Bohmer
Layout design	Stefan Babij
Printed in Singapore	Drummond Printing Pte Ltd

COVER DESIGN - THE TIGER

A Tiger was selected to represent the role CISOs take in a state of cyber battle. The Tiger is a brutal and fast fighter. The Tiger regularly takes on game that is far larger and stronger than the Lion and does such alone. When a Tiger fights, they fight to kill, not just to push aside and get better food. The tiger does not win by intimidation and social challenge, they go into a fight all or nothing. Reference Ariel Williams, www.huffpost.com

I dedicate this book to my Mum who has always been my greatest inspiration, encourager, fan, and the most selfless person in my life. Thank you for all your prayers and your help in making me who I am today. This book is also dedicated to my Dad and my two brothers. You all mean the world to me!

Above all else, I give thanks to the Lord - Psalm 103.

ABOUT THE AUTHOR

Shamane advises C-Suite executives on uplifting their cyber risk and corporate security posture. An influential advocate and business champion for professionals within the cyber risk sector, Shamane actively inspires a fresh outlook that paves the way forward.

Shamane is an international speaker and Founder of Cyber Risk Meetups (www.cyberriskmeetup.com), a platform for security executives to share innovative insights and war stories. Born in Singapore, Shamane now resides in Sydney, Australia.

REVIEWS

"Shamane has put together a great collection of wise insights, with learnings for the most seasoned CISO through to the newest of industry members. She has explored many topics long considered on the fringe of traditional security with great storytelling and insights from industry leaders. It's my hope that by exploring these topics and making them mainstream, we can further strengthen and grow a resilient, humble and ever learning security community."

- Berin Lautenbach
APAC Chief Information Security Officer,
Telstra

"Shamane has done a great job at creating a practical guide for new CISOs or anyone else associated or interested in our world.

Her writing style is easy to read and her content is pragmatic, to the point and insightful. She has even found a way to add humour to a subject that can sometimes be doom and gloom. I particularly enjoyed the section on the top 15 qualities of a good CISO and the "You Did What!" sections throughout the book.

If you invest the time you will thoroughly enjoy the book and definitely learn some lessons along the way."

- Jeff Jacobs,
Chief Security Officer,
Insurance Australia Group

"Shamane has done a fantastic job in bringing together insights from a set of experienced cyber leaders across the globe in this book. The book is easy to read, with a lot of practical examples providing a holistic view of the various perspectives that a CISO role needs to manage on a day to day basis. I thoroughly enjoyed the various war stories and the "You Did What?" sections of the book."

- **Andy Chauhan,**
Chief Information Security Officer,
Ausgrid

"This book is beneficial for both technical and non-technical people, as it highlights the breadth of skillset and traits needed to be a successful leader in security today. This book is a must read for anyone wanting to become a CISO, deals with a CISO/ board or challenging business stakeholders/ executives. The stories in this book by Shamane portray unique insights into the thoughts, challenges, ideas of a CISO or C-suite executive. Her writing captivates the audience throughout each chapter and it is evident that a lot of interviewing has gone into producing this book. I found myself often smiling in agreement, laughing and have taken away some great techniques for the future. This book is beneficial for both technical and non-technical people, as it highlights the breadth of skillset and traits needed to be a successful leader in security today."

- **Jacqui Loustau,**
Founder,
Australian Women in Security Network (AWSN)

"Cybersecurity is a dynamic field and as technology evolves at increasing speed, the amount of data that we're all going to produce, and share is going to rise exponentially. This means the security infrastructure that safeguards this data will come under increasing stress. With experience and insight, Shamane has written a really useful book for existing and aspiring CISOs. It's full of real-world stories from cybersecurity leaders and contains practical strategies for ensuring cybersecurity is built into an organisation's overall risk mitigation program, now and in the future. I loved her unique voice, highly readable style, and wholeheartedly recommend this book."

- Jane Frankland
CEO of Cyber Security Capital, and
Founder of the IN Security Movement (UK)

"A very easy to read and insightful book. Shamane has done significant research in reaching out to so many industry leaders. I particularly enjoyed the real-world examples which make a particular point of view easy to understand. The story telling narrative makes key tips and suggestions easy to remember. A great reference material for both existing and new CISOs."

- Venkatesh Subramaniam,
Global Chief Information Security Officer,
Olam International

"Shamane has captured the experiences and insights of various CISOs in enhancing the cybersecurity of their organisations. These experiences and insights enable us to challenge how we prepare for and respond to the dynamic threats we face, especially as data grows and more systems are utilised. Clearly compliance does not equate to security. Whilst we know the importance of people, process and technology controls, the foremost role of the CISO is creating the right culture. Shamane has brought these aspects to life in her book. I enjoyed the book and its practical approach provided challenges to my thinking and practice."

- **Neil Kenzler**
Chief Risk Officer,
Teachers Mutual Bank

"Shamane reveals the complex and multi-dimensional role of cybersecurity leaders in today's ever-changing business landscape. She revealed remarkable insights from the joint wisdom of security-thought leaders, which will help you, master the art of enterprise cybersecurity. Very thought provoking. I loved her book!"

- **Elik Etzion**
Chief Information Security Officer,
Bank Hapoalim (Israel)

"Cyber has changed enormously as a discipline over the 20+ years I've been in the industry. From being a very technical discipline with a reputation for obscurity more than anything else (except perhaps faceless men in hoodies), it's grown to encompass a range of roles and skill sets which I couldn't have imagined back then. There are many industry trends shaping the growth of the cyber industry but one thing is consistent globally - it's a people business and we don't have nearly enough.

More and more, there is a requirement to share best practice and offer guidance to those in the industry who have had cyber thrust upon them or who have suddenly found themselves needing to understand it in more detail to do their work. Of course, with the workload increasing, there's not enough time spent sharing and networking. Thankfully, Shamane has done some of the work for us all. In her usual engaging style, she's spent time with CISOs and senior security practitioners globally and addressed many of the questions on everybody's lips. She's highlighted some cross vertical and industry wisdom and shown us commonality across the board, which is helpful for those starting out or with established cyber teams.

Her book is a valuable resource for all levels, and she's shared the wisdom in the engaging first person style of a good storyteller. It's very readable and most helpful - whether you want to know what the fuss is about, or build a team to fix a cyber problem. It's a very useful addition to the field of information given the increasing threat landscape and dearth of human capital we face globally."

- Jacqui Kernot
Partner,
Ernst & Young

FOREWORD

By M.K. Palmore,
Retired FBI Assistant Special Agent in Charge,
FBI San Francisco Cyber Branch

In my final assignment for the Federal Bureau of Investigation, I was given an opportunity to see the cybersecurity crisis from an optic rarely seen. As a lifelong security professional, it was troubling to view the difficulty with which enterprises both digest and decide to engage upon the risk associated with a potential cybersecurity breach. It is among the most perplexing business challenges of our time. Over the past several years, I have developed an expertise on the cyber-threat landscape and the challenges it holds for both consumers and business enterprises. This is partly why I was pleased to become acquainted with the author, Shamane Tan, and to share some thoughts about the difficulties information security professionals and other business leaders encounter whilst tackling the cybersecurity threat. We are in the midst of a digital crisis. Unfortunately, from a triage perspective, the crisis continues to present as a slow bleed and is often ignored as an imperative to survival. It is not infrequent that a major enterprise suffers a lapse in controls and protections that makes it the victim of polished and talented cyberattackers.

From my ongoing observations, I have ascertained that several former possibilities are now truisms:

1) Cyberattacks and cyber-related fraud are on the rise and will continue to pivot and target business enterprises

and consumers.

2) Many business enterprises still struggle with identifying viable strategies to combat cyber threats.
3) Information security leaders must continue to view their actions through the lens of risk management to create a resilient response.
4) Cyber-related attacks and frauds will continue until the world's major powers put a stop to them.

Simple, right? When looking to change a culture or implement a strategy in the cybersecurity realm, we often talk about people, processes, and technology. As defenders, we have spent much time on technology development and implementation, some time on process creation, and very little time on the human aspect of the equation. The latter should be the focus of the cyber risk leaders.

A shift is needed. Business leaders must recognize the importance of understanding the threat landscape and make solid determinations about the identification of talent and what it means to create a truly resilient cybersecurity apparatus. I am always amazed when I view the C-suites of technology start-ups in Silicon Valley. The look and feel of the teams are amazingly devoid of diversity. The individuals in these teams are smart, they are young, and they are technically talented with strong or developing business backgrounds. There is usually a technology motivator amongst them who has developed the core technology of the business and each founder is motivated about changing business or the world. Absent from the initial team, however, is an information security professional who is usually not considered essential to the business until the "widget" they are designing becomes the next hot product out of the valley.

Here is the point that they and others seem to miss: the best security apparatus is a mature one. This is covered with some depth and detail in Chapter One: How the CISO was Born. In these mature security apparatuses, the people, processes, and technology have grown with the enterprise. Yes, failure was experienced early on, but the team learnt to excel at mitigation and to develop resilience. The security team was considered vital to constructing a solid business foundation. If only this was the norm!

The scope of job responsibilities for chief security officers (CSOs) and chief information security officers (CISOs) has grown enormously. We have seen the market go from burying the CISO under the chief information officer (CIO), general counsel (GC), or chief financial officer (CFO) to an environment where CSOs and CISOs now have a place at the table and are even frequent presenters to their boards on matters of enterprise risk management.

This is indeed the proper direction in which we should go. The next step is a pivot that is required by the talent recruiters to actively search out diversity and leadership qualities in candidates as well as a technical foundation.

Technology is a tool that can have an immense impact on our fortunes. A good craftsman selects the best tools to craft a masterpiece. Some can even create it with substandard tools, but it is from the vision of the artist that the canvas receives its brushwork.So it should be for the information security realm.

As both a leader and student of information security, it has been somewhat painful to watch enterprises respond to cyber-threat actors. Before we examine grand solutions, we cannot ignore the fundamentals of information security. The majority of investigations that are undertaken by government entities responsible for investigating cyberattacks find violations of the Center for Internet Security's (CIS) 20 Security Controls. Compiled by a dedicated group of global industry professionals, academics, and practitioners, the CIS 20 are the gold standard for the examination of sound information security operating practices within an enterprise, yet they are frequently ignored or implemented in a lax fashion, thereby exposing the enterprise to certain victimhood. I mention the CIS 20 not to proselytize about its efficacy, but to note that in this realm even beginning with or sticking to the basics can be a challenge. To the attacker, it provides a warm sensation of success when the simplest attack vector actually works.

How do we close the gap? There is no shortage of solutions being offered to right the ship or change the current evolution of cyber threat.

We can begin by developing information security leaders. Yes, these are people with a technical foundation, but who also possess a firm understanding of team development and who encourage new approaches and potential failures as a platform for learning. The network enterprise, the cloud, and data analytics are now the norm. Protecting those environments will take more than technologists; it will take leaders.

For a moment, let's examine an often-utilized model for technical start-ups. Some of the most successful ones follow the template described above, but at some point during the evolution of the business, leadership of the new enterprise becomes a consideration. At this point, sound business leaders are sought after to take the fledgling enterprise to the next business level and beyond. The best leaders in this realm are the ones who have not only a sound understanding of the technology, but those who have also practiced the artistry of leading people, especially in difficult times, as elaborated in Chapter Eleven: The Daily CISO - BAU Survival. Today's CSO or CISO is an influencer by the very nature of their tasks and placement in the enterprise. Realizing they cannot simply dictate controls, they are tasked with providing a clear picture for business leaders about the risk to the enterprise and the results of ignoring or partially engaging threats. They are required to strike a balance that provides a suitable level of risk to the enterprise, but also encourages the daily functioning of it. This is a high-wire act of sorts, but there are some success stories to share.

Information security leaders are learning, through experience, business case review, and publications like this one, that there is a need to increase their soft skills in addition to maintaining their technical acumen. How often do you see technical leaders with an eye towards superior communication skills? The answer is increasingly. For example, CSOs and CISOs now adopt the language of risk and speak in terms of likelihood of an occurrence and its impact to business operations. This is the changing world of cyber-risk management.

Shamane has gathered a great amount of wisdom by speaking with more than 30 CxOs from around the globe. This large and diverse group paints an interesting narrative of the state of play in enterprise cyber risk and what it takes to meaningfully engage the aforementioned crisis.

Hopefully, in the near future of enterprise risk management, we will begin to see a world largely staffed by emerging leaders with technical acumen and emboldened with an understanding of the language and function of risks, but who are also astute enough to understand their unique place amongst their peers in the C-suite and those whom they answer to on the board of directors.

CONTENTS

Contents

Contents

Contents

1

CHAPTER ONE

How the CISO was Born

The modern world spins too rapidly. Most of us are caught up with the many facets of our lives and the tides of time and change wash over us. The digital world evolves, too—from the Internet of Things (IoT) to artificial intelligence, machine learning, and blockchain. Yes, I hear you buzzwords bingo people. But the fact remains that as humankind leaps forward in innovation and advancement, there is also less certainty about the world in which we dwell and how we fathom it.

It's hard enough for C-suite executives these days. They must be leaders, who fight the multi-layered cyber battles of business whilst safeguarding the organisation. A study in 2018 by Dr. Michael McGuire showed that the cybercrime economy has grown to at least $1.5 trillion USD each year, and that is a conservative figure. Despite all these breakthroughs in technology, it is hard to keep up.

In the security domain, there's a common saying that the human factor can be the weakest link. I believe, however, that the human element can be our greatest asset. When individuals work together, we become an incredibly powerful team.

I embarked on a journey through different industries around the globe, seeking out C-suite executives—particularly the CISOs—to hear their stories. I wanted to pose questions that they may never have been asked and I wanted candid, truthful replies so we could collectively learn and become stronger.

Here is one sample of the many conversations to follow: During a trip to Singapore, I toured the ICE71 facilities, which is the region's first cybersecurity entrepreneur hub, known as the 'Innovation Cybersecurity Ecosystem at Block71.'

As I walked around, I met a man who introduced himself as Sander Swinkels. Immediately, I noticed he possessed what might be called 'senior gravitas.'

Swinkels says he has two leadership roles. The first, as the CEO of Caecus Ventures, an investment advisory firm and investment partner for national and cybersecurity start-ups and also as the CEO of DutchSec, a highly innovative cybersecurity start-up from The Netherlands.

On top of employing unique techniques and tools to catch hackers, criminals, terrorists, spies, and paedophiles, DutchSec uses an embedded social enterprise model. More than half of the staff has autism-spectrum diagnoses, which also results in their possession of extreme creative skills suitable to this line of business.

THE ORIGINS OF THE CIO

Within the past few decades, at both private-sector endeavours and government agencies, problems arose because IT and business spoke radically different languages and interacted in markedly different ways. Concurrently, there was major spending on new technologies from vendors who promised to radically alter businesses with their beneficial technology. Whilst it is arguable whether they delivered on their Utopian technology, they were definitely often unable to deliver on time or on budget.

Because procurement rules push purchasers toward the best, cheapest option, the industry as a whole kept bid prices low. However, prices were hiked when requests for change came in. It was here that vendors made their margins and those margins were big.

Project costs skyrocketed because organisations were unable to accurately identify and detail their exact needs. Companies could spend a year or more defining their list of requirements—sometimes including thousands of specifications. But no matter how hard a company tried, its list could never be 100 per cent accurate and complete.

Swinkels shared an analogy, "Imagine you are planning your perfect holiday. You look into making the best hotel reservations, or maybe you just need a backpack and a tent. Either way, to make it more fun, you invite a friend or a partner along. You research places you want to visit. However, at the end of the day, something unexpected happens—for example, your companion falls terribly sick during the trip and becomes bedridden. Despite the great effort you put into creating your perfect trip, you're never totally in control when defining what you need to create that perfect outcome."

Out of these dilemmas was born the CIO, tasked with aligning information technology to business needs. Some of the early CIOs were business focused and some were technology focused. No matter which view each held, their key purpose was to reduce budget over expenditure and ensure that the best procurement option was secured.

Purchasers were willing to sign cheques for bigger and well-known vendor brands such as IBM, Oracle, or SAP. These vendors were favoured because of the presumption that a challenging technical requirement would be better met by the products of a large provider. Although they were charged substantially more, the purchaser believed it would get a solution that worked. However, this presumption often turned out to be untrue.

The biggest hurdle for any corporation looking to move away from the legacy ERP and data warehouse-type solutions is that database technology is still based on traditional rows and columns. "But the world as we know it is not so," notes Swinkels. "It's a graph—a network of people, things and events—that defines reality. Every time you abstract data into just rows and columns, you make trade-offs and you have to make choices. You're not capturing all the information; you're simply capturing those objects that you think are relevant to run your business. The redesign process becomes extremely linear."

Due to the excessive overspending by purchasers in the last few decades, procurement rules have now gained extra layers. Instead of enabling the business, these layers became an extra hurdle to procurements and some corporates now end up lagging behind others that are more agile. Businesses and government agencies lag behind because procurement takes too long. Companies with unlimited funds sometimes leap ahead with offensive and disruptive technologies that extend their capabilities. Overall, the speed of implementing and buying lends competitive advantage.

THE RISE OF THE CISO

When did the CISO role first appear? Theo Nassiokas, APAC Cyber and Information Security Officer of Barclays, one of the largest banks in the United Kingdom, states that his role did not exist twenty years ago. System administrators were more predominant. When IT security arrived in the late 1990s, which later became information security, physical documents also became included in the CISO purview. Before CISOs, it was CTOs or the CIOs who managed the additional needs when a company did not have dedicated information security personnel.

Over time, information security was rebranded to cybersecurity, most notably by media and vendors.

"There's no cybersecurity. There's just security. You always need people, process and technology controls," says Nassiokas.

The cyber world requires both cybersecurity and cyber resilience. This has been likened to the debit and credit side of the balance sheet. Cybersecurity is about preventing, detecting and responding to attacks and cyber resilience is how well a company can continue business should an attack occur.

The CISO's biggest challenge is actually similar to that of the previous CIOs. Today, the CISO is overwhelmed by the plenitude of technology solutions on offer, with vendors bombarding them with products.

Nassiokas shares Swinkels' view. "Everyone is selling cyber solutions. Well, from my perspective as an end user, it's not a solution unless it solves a problem that I've defined. Otherwise, it's just a product or a service," he states.

In most cases, the CISO has a small budget, and they may be perceived by the rest of the leadership team as a compliance officer. Unless a major incident happens, they tick the boxes, and report to the regulators or the government that they have done their due diligence when it comes to cybersecurity.

There's a common misconception about compliance: it does not mean that you are secured or safeguarded.

Swinkels shares that most companies he dealt with thought they were looking for a CISO, but in actual fact, were looking for a compliance person. He revealed that he was in an active discussion with an Asian 'unicorn' start-up for a CISO role, but in the end, they mutually decided it was not the best fit. They needed a compliance manager instead of a security leader.

Another CISO told me that when he goes for interviews, companies think they are interviewing him, when in fact, he is interviewing them. He is discovering the problem they're trying to fix. He is figuring out why they are hiring a CISO in the first place. He is delving to discover whether they understand the real role of the CISO and what their security culture is like.

When mapping out an organisation, it becomes clear that a company's landscape is focused on business growth. For example, the goal of a start-up is to grow, formulate new ideas, and invest. They innovate and add new components to their product and service offerings.

The CTO's job is to ensure the business can keep up with the growth by having the platform and the right tools available around the clock. Technology is a way of operationalising the idea, so as to help manage growth. Based on the speed of growth, there is often not enough time to over-engineer or add security layers, and it is essentially iterating scrum.

In such a scenario, the last person in line is the CISO, as the priority is the business growth and putting technology in place to do what is merely good enough. The CISO is viewed as the showstopper and not an enabler. The business is aware that security is important, but that's not their priority for now. Yet, if something goes wrong, the CISO is and will be the first scapegoat.

This is one of the hardest challenges for a CISO. Currently, in many corporations, the CISO is still perceived as the least important person (some have even used the word "annoying") on the leadership team. The CISO ensures that the right risk management policies are in place and that there are ready scenarios and processes for combatting various incidences. Unfortunately, the CISO role and its functions are not prioritised until something goes wrong.

In most situations, working relationships are collaborative and positive. However, in reality there will be conflicting interests with other individuals. A CISO's real interest should be to protect the firm, whilst a CFO's interest, for example, would be to ensure the finances are balanced to best suit the business. The less a CFO spends, the greater the profit and the bigger their bonus. Yet, the more a CISO protects the company, the harder it is to evaluate how safe and secure the company is—hence, the conflict of C-suite roles and why it is vital to have an appropriate balance.

With more popular awareness of major cyber incidents, including when a CEO is fired or massive corporate fines for data breaches, the CISO role will continue to become more and more important.

2

CHAPTER TWO

All CISOs Are Not Equal

How did CISOs get where they are today? To understand the career trajectory of our CISOs, here are examples of their progression toward that role:

- Support Engineer → Security Consultant → GRC Manager → Head of Information Security → CISO
- Security Engineer → Security Architect → Security and Risk Manager → CISO
- Network Engineer → Infrastructure Manager → Head of Infrastructure → Chief Technology Officer → CISO
- Security Analyst → IT Operations Manager → Head of Information/ Operations Director → CTO → CISO

- Protective Security or Police → Head of Security → CISO

Most cybersecurity capabilities fall into one of these four categories:

- Security Operations with an incident and threats focus
- Security Governance and Risk
- Security Product background with a focus on technology
- Security and program delivery

In general, it aids the CISO of a large organisation to have experience in all four domains. Possessing a 360-degree view of the context within which the business operates helps the CISO both affect and influence people.

CISOs with experience in development and infrastructure are in high demand due to their ability to navigate across skillsets and identify the most important allocation of resources. However, as time progresses, the CISOs' backgrounds have become more diverse and less traditional.

Kelly Isikoff, the CISO of one of the world's largest reinsurers, is one such example. Having grown up in an 'FBI family' in Portland, Oregon, she first started her career as a legal clerk for the U.S. Department of Justice and assisted the federal and international investigative agencies in the technical preparation of courtroom exhibits.

Subsequent infrastructure and engineering roles led Isikoff to driving operational efficiency projects for Citigroup Data Office. Her broad infrastructure, engineering, and data science background resulted in a position with Citigroup's Global Security Organisation on the design and buildout of a Cybersecurity Fusion Center. Whilst a part of the

Fusion Center, Kelly led efforts to streamline detection, investigation, response, and recovery capabilities across specialized departments and business units. She was later recruited to work on the buildout of JP Morgan Security Operations functions after a major incident in 2013. As we can see in Kelly's case, her capability falls more closely into the Security Operations category.

Despite the different backgrounds in our CISOs, as we saw in Chapter One, the CISO role is still being defined, and it is important to be aware that not all CISOs are created equal. There is the one man or woman band. They have no security team. They can be the 'CISO' who does everything including approving firewall configuration requests. They are the only person dedicated to security, so they became the appointed 'CISO'. They hold confidence in their title. However, not all CISOs have real power. Some might have been in the same role for more than a decade and there is a culture where the 'CISO' follows everything the board decides. That being said, what the CISO title demonstrates is that the business does recognise that a CISO is needed.

Then, there are the CISOs who are successful in getting the board's backing, along with the required budget allocated and their Security team that they have built up to uplift their organisational security posture. These are the CISOs who have grown into their role and now focus on creating security strategy and solving problems.

UP-CLOSE AND PERSONAL

I asked a few CEOs what traits a CISO should have. The responses included paranoia, genuine concern, and the habit of following major advancements.

"A CISO needs to stay on top on threats and regulatory obligations or requirements and figure out the best technology, whilst still ensuring flexibility and comfort of today's workforce and sensitivity of assets," said Uzair Gadit, who is based in Hong Kong. Known as a serial entrepreneur, Gadit has extensive experience in creating scalable platforms for cybersecurity, the cloud, and the gaming industry.

Carmen Marsh, CEO of Inteligenca Inc. a global consulting firm in California, listed these points:

1. A CISO puts a company's security and privacy first and provides the latest knowledge of threats, with solutions recommendations and total cost of ownership (TCO).
2. A CISO is a company's champion for raising awareness of cyber risks across the company.
3. CISOs work collaboratively with the rest of the executive team.
4. CISOs hire, mentor and retain the best diverse cybersecurity talent to protect their companies 24/7.

Mike Webb, CIO of Commonwealth Treasury, told me, "CISOs get it wrong when they don't understand the organisational context. The best CISOs I've worked with can talk in business terms, negotiate, and champion a culture that incorporates and embraces security."

Commonwealth Treasury is best known for producing the Australian Government's annual budget, but they also work with the states and territories on significant public policy matters, analyse global and domestic economic conditions, and shape the tax, superannuation, and financial systems that are fundamental to the function of the Australian economy.

Webb places a premium on CISOs who see their role as being inseparably interconnected with the business as a whole rather than simply focusing on traditional cybersecurity. If a CISO understands business objectives and how to meet them securely, it negates the behaviour of anyone seeking ways around security controls. It promotes a self-supporting culture that shares problems with all executives rather than just the security function.

CHAPTER THREE

3

IT CISOs Vs. Business CISOs

McAfee's APAC CTO Ian Yip alluded to the term "VUCA," in describing most business environments today where things are in a constant state of flux - being Volatile, Uncertain, Complex, and Ambiguous. VUCA has, unfortunately, become the normal working environment.

Those who come from a stable, static history where change is infrequent empathise with just how VUCA things now are. To function well and to do a job properly, people must let go of old ways, as well as not rely on resources that have always been there, for they may be gone tomorrow.

A business CISO gets engaged across all aspects of

the organisation such as the legal, compliance, product, sales, facilities (physical security), finance and even the HR teams. They all have different security needs. For those trying to become better at their jobs or step into the role of the CISO, some of the key qualities that a business CISO should have include good communication skills, empathy and resourcefulness. Yip unpacked them for me in detail.

#Good Communication

"We become better communicators by staying specific and not overly detailed. We better communicate cyber and technical risks by using the stakeholder's language. To do so, we must understand how they measure success," Yip says.

"Take, for example, the risk of a data breach. It is important to know the most likely ways an organisation can be compromised, as well as what the resulting impacts could be. These might include reputational damage or a regulatory fine. In the case of a General Data Protection Regulation (GDPR) breach, exposed revenue could result," he states.

"You can't avoid the real impact," Yip notes. "There is always some fear at the enterprise risk level. It doesn't just apply to cyber. This is similar in the context of safety risk, albeit that impact can be fatal. There is no way to sugar coat this.

"The point," Yips says, "is to articulate the risk in a language that the business and a non-security person can understand. Show them how the risk can be addressed, how it can be measured and how it can be mitigated."

#Empathy

Why is empathy so important for a CISO position? In the later chapters, Yip talks about the arrogant experienced professional — the one who thinks he is the smartest person in the room. One can never have a constructive debate with someone who thinks they know everything. Arrogant people also tend to lack empathy. They naturally find it challenging to manage stakeholders or influence others in their organisation unless that culture is a dictatorship—a model that is ultimately unsustainable.

The difference between leaders and administrative managers is that leaders understand they don't have all the answers. Yet, they innately know how to manage people, stakeholders, and teams. Administrative managers, on the other hand, assume everything they say will happen. They lack enough empathy to understand how their behaviour affects those around them.

To soften someone in order to influence them with the facts you need to present, it is essential to connect with them at a more personal level. For example, find a connection point that allows an informal conversation. In this way, walls may be dropped to expose the real human, rather than the professional who thinks he knows everything.

#Resourcefulness

Resourcefulness is a core skill for CISOs. Often, clear instructions are not laid out. CISOs must often figure out how to get things done on their own.

Even in the most structured environments, senior

executives need to be enterprising. Yip said of the people he has worked with, "The ones who get the most things done are the most resourceful. They can figure things out, even if they are not sure—no matter the challenges, the uncertainty, complexity, volatility, and ambiguousness of the task."

Resourcefulness means understanding what is at your disposal that you might not think you can use, but you can. Resourcefulness also means asking yourself, "Whom can I call? What can I read? What can I try? Can I bend this process a little? Can I exit the process and not get into trouble?"

WHAT IT MEANS TO BE ACCOUNTABLE

I asked the CISO of the Western Australia Police Force his biggest fear. His answer was that in the event of a calamity, the organisation would not be able to successfully implement disaster recovery plans. I could understand how gripping that fear could be, wondering whether your actions, or lack of action, could result in the domino crash of the state's well being.

Another CISO highlighted a regulatory trend now in Hong Kong and creeping to Singapore, Australia, and other locations. 'Whilst there have been cases of heads of securities trading or other instrumentals – not yet cybersecurity – being personally fined by regulators for inappropriate conduct, the Senior Managers Regime (SMR), which first came into effect in the United Kingdom, takes it a dramatic step farther. CISOs or heads of security have been included in the regulations under "Senior Management Functions," thus deeming these professionals potentially culpable for poor decisions as executive directors.

This means that APAC CISOs, under their current

employers, are accountable to the security and banking regulators in Hong Kong, Singapore, and other locations. The regulators possess their personal information (name, date of birth, home address, phone numbers, and more). The CISOs are personally accountable, with the real potential of personal liability looming over them.

Since personal accountability may also translate into personal legal liability, it is important for CISOs to be transparent and accountable, and to maintain evidence of oversight to show that "reasonable steps" have been taken to prevent, cease, or remedy breaches.

A CISO must be a dynamic lateral thinker, not a linear process follower. Being a good CISO is first about people, then process, and finally technology.

CHAPTER FOUR

4

Into the Future

Although many CISOs come from a technical background and possess an extensive understanding of cyberattacks and threat intelligence, their role closely follows the trends of business and market culture.

#Less on Technology

CISOs see the future of their role in different ways.

One group feels there's an increasing appetite for modern CISOs with strong technical backgrounds, but who also communicate risks in a non-technical manner to business professionals.

"A key driver for many of today's CISOs is to build security into operational efficiencies across engineering and infrastructure disciplines, whilst modernizing on-premises security solutions, Kelly Lindsey Isikoff, CISO, RenaissanceRe, told me before.

'Increasingly, CISOs do not simply need to understand technical architecture or how a firewall works. Instead, they need to understand the role technology performs in the corporation. There are teams with the expertise to handle everything else. As agile development and cloud-hosted environments have become key transformational drivers of traditional IT departments, modern CISOs need to articulate business risk with defence-in-depth options. It is common consensus amongst CISOs that the old command and control management style does not work. Security needs to enable business IT transformation. At the same time, a new challenge has arisen with the increase in compliance requirements around data and breach management. There is now a heavier focus on reporting governance and compliance.

Jeff Jacobs, CSO at the Insurance Australia Group, foresees the likely convergence inside progressive companies of cyber and physical security, privacy protection, and fraud prevention. There are already organisations, including IAG, that bring together cyber and physical security.

"I suspect additional convergence will happen for many companies, especially those in financial services. I have the view that the CISO will likely morph into the CSO—as I did—then potentially morph into more of a chief security and trust officer. We might even see other risk classes blend with the security function at some point," Jacobs commented.

#People Management

The other group predicts that as budgets for security and governance increase, teams will continue to grow. Thus, the role of the CISO will become more about team management, as well as the soft skills mentioned earlier and to present complex subjects to the board and budget management.

"Hopefully there will be fewer guys in this role as well!" Ian Yip, the APAC CTO, chimed in. "At the risk of stereotyping, a lot of the arrogant personalities tend to be males. Females are a lot more empathetic and open to discussion and continuous learning. I find that the best CISOs are the ones who can communicate, are empathetic and can lead. They are neither the smartest persons in the room, nor the ones who think they are managers, nor a dictator, and not the most technical people in the room, either."

Frances Bouzo, CISO at iCare, one of the largest insurance services in New South Wales, Australia, which protects the state's most iconic assets, including the Sydney Opera House and the Sydney Harbour Bridge, shared her experiences from 20 years experience in the technology industry. "I'm very passionate about encouraging more women to join this growing field. Cybersecurity may seem a challenging field to work in but like many other complex fields, diversity and diversity of ideas are key."

Bouzo highlighted the importance of having the confidence to step through any doors of opportunities in cybersecurity and embrace the challenge. "Remember, often others see in you what you don't see in yourself. If things get tough, don't doubt yourself. Remind yourself of your purpose, remain confident, and push on."

She encourages women to speak to other women in the field, expand their network and knowledge, adopt a lifetime of learning approach and focus on their strengths to help them succeed. "Every step women take to improve diversity in the field of cybersecurity creates opportunities for those coming behind. Don't forget that you can also pave the way for other women."

Dan Lohrmann, the former Michigan State Government CTO who later became a CSO, shared that there was a good mix in their Michigan cyber teams, with both genders represented as well as many ethnic backgrounds. "I found that women often brought fresh insights and ideas." He commented. "We were fortunate to have a great team that people wanted to be a part of, so we had the choice of the best and brightest, which is very hard in government today."

Dan offered this advice for security leaders: "Bring in young professionals from diverse backgrounds as interns or students to ensure a diverse culture and encourage innovation and learning."

#Strategic Function

The third group is adamant that reporting structures will also change. Most strategic companies that put security as a priority, albeit not always first, will have the CISO report to the CEO, chief risk officer (CRO), or CFO. The most progressive will have the CISO report directly to the CEO. These are businesses that understand security as a strategic function and back it up with their operating model.

Although reporting models are a visual representation of hierarchy, the best functional models don't rely on them.

During one of my chats with Yip, he noted that companies that rely on reporting models "take hierarchy literally; the CIO holds the power and the CISO doesn't. In those cases, that's where you need hierarchy to reflect authority."

In a well-oiled corporate reporting structure, no matter where the CISO sits, if security is viewed as strategic, the CISO should be able to go straight to the CEO without causing offense to anyone. However, in most cases, it would still require an actual reporting model to force people to shift their thinking. Most CISOs agree that as there is no distinction between ebusiness and traditional business, in the future, there shouldn't be a distinction between cybersecurity and security. The CxO role should encompass all aspects of security.

I wanted a CEO's opinion and reached out to Uzair Gadit. PureVPN, a brainchild of Gadit's, steers the information security wing of all the ventures of the company Disrupt. Time and again, Uzair has proven himself to be ahead of the curve and has led from the front.

What is his view on the future of the role of CISOs?

"I think it's going to be a hybrid role. Given the growing surface area that CISOs will be expected to cover in future, it would be impossible for a person or a department to do it on their own. It will have to be done in-house, crowdsourced and outsourced. It will be an interesting hybrid that way," Gadit stated.

Israel Baron is Israel Railways' first CISO. He declared that the role of the CISO needs to be as high up the management hierarchy as possible, under the CEO or a direct deputy. Ten years from now, he said, the CISO is going to be a prominent figure in every organisation. This is already happening, with the shift from a business battlefield to the cyberspace battlefield.

Masked terrorist attacks can already be conducted in the cyber world without the attackers moving from their chairs. There are just too many scary scenarios—from the derailing of trains to nationwide loss of power to fiery explosions in a motorway tunnel full of heavy traffic.

"I'm not afraid of regular attacks such as DDoS," Baron told me. "Despite the discomfort, we will manage it. What I'm afraid of are attacks that will cause trains to collide. I don't see critical infrastructure as commoditised; my family rides the Israel Railways."

5

CHAPTER FIVE

The Good CISO (Part I)

Albert Einstein's definition of insanity is doing the same thing over and over again and expecting different results, right? Well, this statement wasn't his. We are the insane ones who repeatedly attribute it to Einstein. That said, however, I've held the wisdom closely to me for many years. I believe it is incredibly important to actively seek out mentors, to look for creative ways of doing things, and to learn from others' mistakes.

Now that we have explored how CISOs came to be, and also looked at how the role is expected to evolve during the next few years, we will look at what it takes to be a good CISO in the context of constant change.

I asked more than 30 CISOs to share their thoughts with fellow security professionals, up-and-coming CISOs, and those seeking success in this industry. This resulted in my list of the Top 15 qualities that came up in conversation.

#The Top 15

1. *Fully understand the role you are seeking.* Often, people who are promoted end up hating their new position. Ladder climbers need to understand what they will give and get from the role, rather than merely questing for a title.

2. *Establish yourself as subject matter expert who contributes to the success of the company.* A good CISO is also knowledgeable in the areas of engineering, networking, infrastructure, and risk management. Their clear, simple style helps nurture relationships with the executive team.

3. *Prepare to be wrong and to admit it.* Nobody knows everything and a fancy title won't help.

4. *Accept being the bearer of bad news.* It's what CISOs are paid to do.

5. *Consider security as an enabler.* Its primary purpose is to improve business performance.

6. *Stay on the top of the latest technologies.* CISOs demonstrate their value by protecting company data, as well as implementing the latest available safeguards.

7. *Be a trusted advisor.* CISOs let the board make decisions on risk that is acceptable or the need to allocate more budget towards reducing risk.

"Manage risks, says Shane Moffitt, assistant CISO to the Victorian Government. "Don't just sit around admiring them."

8. *Offer advice and decisions that are based on the organisation's objectives.* CISOs listen and learn from the business's top concerns.

9. *Seek executive support.* A supportive CEO makes all the difference to the success of the CISO.

Dan Lohrmann, CSO and security mentor, states, "A commitment to excellence starts with the highest-level executives. Having a strong endorsement from top elected officials is great, but sadly is not always the case. If you can't get your top leader to vocally support cybersecurity, try to find business-side champions to help shape your message. Most organisations have leaders, followers and laggards on tech, so start with a 'coalition of the willing' who support your efforts to get some needed momentum."

"Sometimes, you've got to be like a dog with a bone,"

adds Darren Shearsby, CISO, Compass Group Australia, the largest contract foodservice company in the world.

10. *Establish a good relationship with the board.* Good CISOs will find a way to earn trust with key business stakeholders.

"If you become a CISO, and your job title is a CISO, but without a good and trusted relationship with the Board, it's like having a gun with no bullets," says Chin Kiat Chim, CISO of Global CISO at DHL Express, the world's largest logistics company.

11. *Don't focus exclusively on IT security.* Good CISOs can't forget about personnel and physical security.

Carmen Marsh, CEO, Inteligenca Inc., states, "My advice to the CISOs of today is that the position might not be seen as important as it should be, but you need to find a way onto every executive roundtable and into every boardroom because you are critical to the success of the company. Cybersecurity and privacy should be part of the company's operation from the budget, business resilience, and culture perspective."

12. *Be prepared to be short of staff.* Good CISOs use cyber trustees.

Baron, formerly of Israel Railways, shared how he personally grew a team of cyber trustees when he had zero resources. "I identified key representatives from different teams and trained them in cybersecurity. It made all the difference that I had a very supportive CEO. He agreed to endorse all those who passed the cybersecurity training with a diploma. This declares that on top of their current position, they are nominated to be a cyber trustee. The communication is clear. Everyone in their department knows that they are the cyber trustees. They act as my point of contact for everything. I get calls from them if there's a new project that security needs to be involved. They work together with me."

13. ***Educate the management and employees about the threats.*** In this way, CISOs pull everyone together to make business as resilient as possible.

14. ***Addressing the human factor is unavoidable.*** Good CISOs win by creating awareness programs using real-life scenario-based drills.

15. ***Perseverance and determination are needed for cyber defenders to succeed.***

"Setbacks can be disheartening," says Dan Lohrmann, former CSO of the U.S. State of Michigan, who once worked at the U.S. National Security Agency (NSA). "It takes ongoing vigilance to be successful, and in cyberspace, the online attacks never stop. Even if you are successful for a long time, what worked last year may not work this year. The strategic and tactical battle is constantly evolving."

6

CHAPTER SIX

The Good CISO (Part II)

In this chapter, CISOs explore in-depth the key qualities they think other good CISOs need, including practical knowledge, the proper mindset, leadership abilities, humility, good mental health, and knowing their purpose.

#Practical Knowledge

One CISO defined the biggest problem in cybersecurity as the lack of practical cyber or information technology knowledge among a majority of the senior security leadership. Most of these leaders come from a risk and compliance background, which was appropriate ten years ago but not today.

Today, businesses face then-unimaginable cyberthreats, and cyberattacks that require leaders' technical and operational understanding to create pragmatic defences. It is important for all leaders to increase their skills and technical understanding of the complexities a cyberattack can involve. This knowledge must range from the simplistic, such as not clicking links and identifying suspicious emails, through to complex red-teaming and penetration-testing exercises.

As part of a cyber defence or security posture, a CISO should understand what and where the risks are, as well as the granular, such as the exact phase of the cyber kill chain whereat an attack should be stopped. CISOs must also have practical knowledge of the environment and must also understand the complexity of network architecture, operating systems, shared services, applications, and users.

Unfortunately, a number of CISOs find this understanding deeply lacking in many of their peers. This cripples the definition and communication of the most effective solutions and the resulting beneficial value it can contribute.

#Mindset

Despite having preventative controls in place, including firewalls, intruder prevention and detection, and other solutions, what happens if an organisation is still breached? Where is the last line of defense? How strong is the business continuity program? The issue is not just about ensuring that IT security is looked after but also building business resiliency.

To do so, the total picture must be seen—from the

organisation, down to the department or business unit, then to the teams. Viewing technology in a silo will not effectively secure the entire organisation.

As on a battlefield, prior hands-on experience helps, but each battle can be completely different and therefore challenging. Ultimately, it's all about mindset; CISOs don't need to be experts in the different security technologies. A CISO's expertise should be in defining the problem, formulating the right strategy, and recognising a commercially feasible and sustainable solution.

Hai Tran, CISO of the Western Australia Police Force, believes that having diverse experience in both business and cybersecurity across different sectors gave him an edge. He was able to combine lessons learnt from the other sectors. In one of the conversations he had with me, he shared that one of the key qualities of a good CISO is being able to put on the CEO's hat to understand that security is about enabling and enhancing business performance.

Theo Nassiokas of Barclays said that despite the different roles he has had over the years, which included listening in on and deciphering foreign-language telephone intercepts, his work principles have always remained the same. His passion in protective security as a police officer gave him a unique security mindset in the corporate world, whether as a banking employee or as a private sector consultant.

#Leadership

In the earlier chapter, we covered the importance of people-management skills. Good people-management skills have helped some CISOs get where they are today. However, it's not just about excelling in stakeholder management but inspiring security awareness in people who originally had no understanding of the elements of security.

"A mark of good leadership is the ability to find good people and to help them reach their potential," notes Trevor Cushen, one of Australia's largest payment provider.

As one CISO specifically emphasised, it is crucial not to belittle your breadth of experience. Any experience, no matter how narrow, can provide wisdom to draw on. In the meantime, aspire to strengthen your leadership skills whilst in the journey of gaining new experiences.

#Humility

Those in mid-career, take heart: this segment is encouraging. Ian Yip has worked with a number of CxOs and offered his observation about working with people who are set in their ways. "Let me give you an example," he offered. "If I were to see a doctor, I would rather see one who is in the middle of their career, not someone who is the most experienced doctor in the world," he told me.

"Doctors who are extremely experienced, who have been around forever, are used to being the smartest person in the room, and along with that can sometimes come arrogance and the absence of self-awareness and humility. This means that, lacking these attributes, they will always think they are right. In reality, a number of things they've learnt during their career might now be wrong because of the speed at which science moves. They will walk around assuming everything they've ever learnt remains true. Now, someone in the middle of their career would hopefully be more humble and up to date with the latest medical developments and they would provide me with the best care."

My conversation with Yip made me wonder whether our younger professionals have been trained just to pass final examinations or do they recognise that learning is a continuous journey? How are we academically educating our future CISOs?

"If you are talking about people who come from a technical industry, they tend to be fairly intelligent, as they often need to be, working in technology and cybersecurity. By the time they get to a very senior position, the same analogy as the doctor holds," Yip said.

Dan Lohrmann, CISO of the State of Michigan, despite believing that Michigan's cybersecurity was the best amongst state and local governments, they kept pushing forward. Whilst staying abreast of cutting-edge cyberthreats and stealth cyberattacks, they made sure they knew their strengths and weaknesses. There was no assumption that they know everything. The State constantly sought out best cybersecurity practices from around the world to benchmark against.

#Mental Health

With the work that our CISOs are doing on the front lines, the CISO job can be incredibly stressful. It is not uncommon to often face situational emotional highs and lows. One of the questions I posed to the CISOs was this: How do they keep pressing on?

I received a variety of answers:

- "Keep a 'head above water' approach to help you get through your crisis."

- "Drink heavily! No, seriously, the trick is to accept you can't always know if you are doing the right things. It's good to have independent assessments to sanity check your work," said this Irish CISO, who—despite his current responsibilities—manages to retain his wicked sense of humour.
- "Give support to your team and allow them to challenge things to keep you on level ground."
- "Keep things realistic."
- "Look forward to Friday afternoons! And the weekends. And holidays! All right, I'm just passionate about what I do because it's how I'm wired," said this CISO, who even makes time to train in Kali martial arts every week.

How do they keep up with everything that's happening? It's a constant juggling act, CISOs confided.

Most of the CISOs I spoke to expressed that cybersecurity is one of the most difficult jobs in the world, as it includes security, networks, IT systems, user experience, people, culture, applications, processes, compliance, and more. Additionally, the purpose of the CISO role is not yet clearly defined.

One CISO put it: "When things go well, no one appreciates you. When shit happens, you get blamed."

Cybersecurity, the CISOs state, is not for the fainthearted.

#Knowing Your Purpose

Chin Kiat Chim, whom I met in Singapore, got even more real with me. He compared the two industries of his current role with DHL and previous role with Shell, an oil and gas company.

The oil and gas industry is heavily regulated, and very

process-oriented. They are well controlled and centralised, as they are also heavily audited.

Each person at Shell is an expert in his or her domain. The challenge, however, is that everyone has such vertical depth that they may lack horizontal foresight, making them blind to the bigger picture of risk or opportunities outside their role.

The logistics industry, on the other hand, is just the opposite. It is decentralised by geographical operations yet requires working as one global network to deliver fast, reliable, and superior service. For Chin Kiat, his new role with DHL took two years to settle into. Every day involved reshaping himself to maximise his contribution to the company.

He said what kept him going was this thought, "Even if I were to quit, I would still face the same challenges elsewhere." He had to acknowledge that he could not apply everything that he learnt from Shell to DHL. The culture and the way of doing the business are too different.

He decided that instead of looking for new opportunities or an excuse to throw in the towel, it would be a better use of his energy and time to truly understand his company. For instance, at a high level, as he gained familiarity with the business strategy and company culture, enabling him to identify key stakeholders and learn their communication styles so that he could gain support for the security program.

From a profit and loss perspective, the company's bottom line (i.e. profit) and business pain points were insights to how security initiatives could add value to the most profitable business functions. From there he was able to find potential opportunities to enable business growth.

When he first entered DHL, he was brought in to fill the

IT security and compliance role. He had to build up security capability almost from scratch. Instead of complaining about it, he thought it would be more fruitful to think positively about what he could achieve. He shared a few pragmatic approaches that I will reveal in the later chapters. This mindset helped him make a positive business impact, and in doing so, gained recognition from the business as he was moved very quickly into the CISO role.

What motivated him were three things: the new things he could learn, his conviction to improve security maturity and culture, and the ways he could add value to the company through his success. Chin Kiat is someone who genuinely enjoys challenges. He likes dealing with the unknown and is constantly trying to find creative or innovative ways to solve problems. For him, it was not just a battle between the good and the bad guys in the cyberspace. It was a matter of who was going to win this war that ensured the long-term success and business resiliency of DHL.

At the end of the day, one needs determination to find their purpose. If not, it will be hard to outlast the race.

7

CHAPTER SEVEN

The Good CISO (Part III)

In this chapter, we will make a worthwhile exploration of CISOs' motivational drivers for CISOs. To solicit their, insight, I asked them to talk me through what they would consider to be a great day.

Trevor Cushen answered, "Thinking about this, it always comes back to days when I've helped people on my team. Getting someone promoted or organising a cool business trip for them is great, but really it's about seeing people succeed and having helped in some way for that to happen."

Kelly Isikoff of RenaissanceRe said, her best day was watching a team work "that loves what it is doing and is learning and improving. Seeing that is one of the most rewarding feelings I've had as a CISO. Security is everyone's responsibility and I find educating staff about how to be safe online very rewarding."

Theo Nassiokas of Blarclays said, "The biggest buzz for me is when I succeed in complex projects where people think that a fix is unlikely and the odds are against me."

One CISO shared a hypothetical example with the following disclaimer: Any resemblance to any living or real persons was purely coincidental.

A regulator became angry because a bank had seven core banking outages in as many weeks that hadn't been reported to regulators.

This CISO decided to undertake a root-cause analysis despite it not being part of a business-as-usual problem-management process in technology teams. This was a customised and independent assessment done on behalf of the regulator to determine the root cause of the core system outages and why those outages occurred, with a view toward mitigating any recurrence. Secondly, the assessment sought to discover why the bank failed to notify the regulator of the outages within the prescribed timeline.

The regulator had to approve the independent assessor. This means that they had to endorse the assessor's qualifications and experience, as well as that their reporting line was sufficiently independent from the technology function where the technology failure occurred.

The most challenging part of it was to deliver a sufficiently independent and evidenced assessment to the regulator to give it a high level of assurance that the problems had been clearly understood and mitigated, thereby avoiding regulatory censure that would have caused reputational damage to the bank, amongst other issues.

All the above is a difficult feat to accomplish, the CISO opined, but it is precisely the challenge of overcoming such odds that keeps him motivated.

Another CISO concluded that his best day was when he finally gained the trust of his board and executive peers after tirelessly working on each individual for months. "Now they see me as a valuable partner who enables them to be successful in their role, and that gives me the biggest satisfaction—to know that I have done my job well."

Ian Yip revealed, "It's being able to work in an environment where I feel my actions and my team's actions make a real impact on people and organisations."

It is heartening that this incredible group of leaders bravely venture into new territories and lead from the front whilst protecting the world as we know it.

YOU DID WHAT?

I asked CISOs if they had any interesting stories to share with readers, and they definitely do.

One CISO shared this experience, "We asked our users to choose a password that contains at least three random words." He paused for effect then continued, "During a password audit, we found a user who had the password 'threerandomwords.'"

* * *

By the time professionals reach the level of CISO, they are usually dignified, distinguished, and polished. Yet theirs is a role in which anything can happen. From a breach, to a cyber- or state-sponsored attack, the CISO's job can be stressful.

Behind every experienced CISO is a real person, who is doing the best that he or she can. But at the end of the day, we are all mere humans.

Our next story features a head of security who confided, "If you think that presenting to a board is bad, it can't be as bad as falling asleep in front of the customer who is right across the table!" he laughed.

"Tell me more," I encouraged.

With a twinkle in his eye, he continued, "I fell asleep right in front of the customer who was busy talking directly to me. My colleague woke me up to answer the question I hadn't heard."

"Did you snore?" I asked.

"I'm not sure if I snored! I was working with a European bank, which was the customer, and which has a presence in six different countries, so I was travelling to a country each week. I was four to five weeks into this when I felt a flu coming on. But when you've had all your flights and hotels booked, you can't just take a sick day, so I had to just get on with it!"

He smiled and continued. "Luckily, I'd mentioned I wasn't feeling well at the outset, so when I dozed off I had an excuse. But it was still extremely embarrassing."

When asked how the customer took the unscheduled nap, he said, "They were fine and a bit concerned. They even offered me the company doctor! I continued for a few hours more but ended the day a bit earlier. When I got back to the hotel, I just slept 12 hours straight right through to the next morning. The next day, I picked up where I left off… and stayed awake this time!"

CHAPTER EIGHT

8

CISO and the Business: Power, Money, and Influence (Part I)

We all know that CISOs have their job cut out for them. In this chapter, I ask CISOs to list their biggest challenges in dealing with boards.

One of the biggest challenges is for CISOs to present complex subjects to board members in a short and concise manner because the members aren't knowledgeable in technical and cybersecurity matters.

Although cyber issues are not new, they are becoming much more prevalent at the board level. Moving forward, it will become essential for CISOs to master communicating security risk in a language and manner that the boards understand, translating the CISOs' language into the Board's, focusing on the business priorities and the risk currently being faced, as well as the risk if nothing is done to address it.

Once a CISO gets the message across successfully to the board, establishing the importance of cybersecurity initiatives, the next challenge is to make them see it as a business priority. This requires a mix of both hard and soft skills, as CISOs must complement their subject matter expertise with the subtleties of communicating to those from business backgrounds.

For a CISO in the Asia-Pacific region (APAC), the challenges are different. At the APAC level, complexity does not center on making strategic technology decisions that go to corporate headquarters, often based in the UK or USA. The complexity is in dealing with a large number of regulators in a relatively small area. By area size, APAC is bigger than the USA or Australia, yet amongst the 16 countries from the North to Southeast Asia (SEA) there are more than 30 active banking regulators for the finance industry alone.

It is complex enough to comply with the regulations of all the countries. Imagine needing to bring those local regulatory requirements to global management for consideration in its strategy—that truly adds another layer of complexity. Unless you've worked in a similar role previously, global headquarters executive can find it all but impossible to understand the APAC region's regulatory concentration.

To top it off, as more and more outsourcing of security functions takes place, third-party risk and management will grow in importance. I asked a former NSA employee how the agency does things differently. He said that NSA certification programs focus on verifying skills and experience. Their training is world-class and they emphasize the importance of testing and interoperability. This includes managing vendors with detailed checklists.

Another common challenge faced by the CISO is justifying budget requirements and multiple changes due to a shifting business and digital landscape. Each day, the cyber risks grow as more data and systems are utilised and relied on by business.

DEALING WITH POWER

Theo Nassiokas, whom we met earlier, has worked in seven banks during the last two decades. He also ran his own consultancy, and prior to that, was in public service.

He says that the most common of CISOs' frustrations is the knock-back on their requests for funding due to other priorities. "Then, when something goes wrong, or there is a major breach, it's the CISO who gets the blame," he adds. "The board is not going to say, 'We don't blame you because we didn't give you funding.' The board is going to say, the cyberbreach happened while you are CISO. You are accountable; you are obviously incompetent. You are out."

Isn't that brutal? With so much resting on the shoulders of CISOs, it is indeed sad to note that some companies have used their CISOs as scapegoats. This is why influencing upwards is so critical.

Nassiokas addressed plausible deniability—the ability to deny knowledge of, or responsibility for, actions committed by members of an organisational hierarchy because of a lack of evidence, even if they were personally involved in or at least wilfully ignorant of the actions.

In times of crisis, it is not uncommon for CISOs to hear

that they should have warned the board that a breach or an incident would happen and that the board would have given over the funding to fixed things before it reached a crisis, if only it had known. This is, of course, an easy statement to make in the aftermath. CISOs must become incredibly frustrated by such responses when they have in fact warned the boards of the danger right from the start.

One method by which a CISO can deal with plausible deniability is to document risk assessment or threat analysis recommendations. When asking for funding to mitigate risks in a commercially feasible way, Nassiokas sent that document to his board. However, the tricky part is to consider the stakeholders and be politically savvy.

There are layers of complexities to this. On one hand, when an issue is documented and put it into a corporate email system, it becomes legally discoverable. Whilst some stakeholders might not mind this after things are explained to them, there are those who would not, in any circumstance, want to receive risk assessments or threat analysis as justifications for funding. This places them in an awkward position where funding has been clearly justified but not provided. (It is also important to note that there are also things a CISO might not want to commit to writing.).

Documentation can stave off plausible deniability if your employer decides not to fund risk mitigation: They can't say they didn't know because they were not told. Yet, this should not be used as an easy excuse to drive funding from a perspective of regulatory compliance. Security should instead be driven from the quest for the level of risk mitigations that make sense as a cost of doing business.

For example, the entire world is online and it's a foregone

conclusion that companies will be hacked. It is only a matter of when, how often, and how much it will hurt. If a company wants to avoid being hacked, it must literally shut down its business. Therefore, solutions need to be commercially feasible. It makes no sense to spend a million dollars to protect a hundred thousand. Yet, it may be responsible to spend fifty thousand dollars to protect that one million. It's about perspective and context.

Chin Kiat, the CISO at DHL Express, said, "We are constantly spending money, but we don't see our cybersecurity maturity or resiliency being improved!" is a refrain CISOs often hear from boards that think compliance equals security. Therein lies the gap from a strategic perspective, as it results in differences in implementation.

Later, in the next chapter, we will further uncover the style in which Chin communicates with the board.

THE ART OF BUSINESS LANGUAGE

Whilst sometimes it takes scare tactics to make a business case or sales pitch, the focus should be on presenting a clear benefit for the cost. CISOs are helped do this by getting people on their side in advance and rallying champions to their request.

"As a business, you have to maintain positive relationships and trust with your clients or customers. I know of a few examples where companies went out of business because of a data breach becoming public knowledge. You don't want to be in the news as one of those companies. As a business, you must appear strong and resilient to cybercrime to gain and retain the client's trust," Carmen Marsh of Inteligenca told me.

There are practical tricks that CISOs I've spoken to use in

preparation for board meetings. For example, some meet with board members in advance in a more informal setting. This approach has proven helpful in increasing board members level of understanding.

Another tip is to have a document packs delivered in advance of board meetings that clearly explain the proposed plans, actions, results, and the cost impact.

A CTO of a multinational company, shared that the fewer the people in the meeting, the more productive it will probably be. However, that is not always in the CISO's control. CISOs must do their homework by knowing whom they'll be talking with and what their backgrounds are. In as much as possible, they should search for their audience's preconceived notions or areas where they may be agreeable or confrontational.

I quizzed some of the CISOs on presenting to their boards. Each has different techniques, but it was interesting to note the patterns in their answers.

#CISO Techniques

One CISO believes that the board should be presented with information that allows them to choose risk treatment options rather than presenting a general story of doom. For example, take the scenario of deciding how frequently systems should be patched. The CISO said he would first present the board with the monthly, quarterly, biannual, and annual costs of patching. Thereafter, he would present the benefits and risks of each option, then let the members choose.

I also conversed with Dan Lohrmann, who served on the Executive Board of the Multi-State Information Sharing and Analysis Center (MS-ISAC). MS-ISAC is focused on sharing and coordinating cybersecurity information between state governments, federal agencies, and centers such as the National Cybersecurity and Communications Integration Center (NCCIC) at the U.S. Department of Homeland Security (DHS).

I asked what drives him to allocate budget to the different initiatives proposed by the 50 U.S. states, he spoke against "FUD—fear, uncertainty and doubt." FUD alone won't bring you more resources, he said.

During a presentation, news of a recent data breach or ransomware attack can sometimes be used like an appetizer before the cyber main course. However, he said he learnt long ago that security professionals would fail if they always brought bad news.

One cybersecurity executive took to LinkedIn to write about his aggravations after hearing another presentation focused on how bad things are in cyberspace and the attendant challenges to security leadership. He wanted more presented than headlines and statistics from Google, and he wanted the communication kept positive. Bring solutions, not problems, he said, adding that CISOs should clearly communicate using business language on how to make improvements.

Another CISO told me that he would focus on presenting a clear strategy via a document pack sent before the meeting, as well a brief slide deck summary. The focus, he stated, is on remaining concise and clear. He stresses that the board has accountability in protecting the company, so it is crucial for him to provide assurance that the right things have been done. External independent assessments can provide this assurance as well as allow CISOs to benchmark their own maturity within industry frameworks and in comparison with their peers.

CISOs should remove technical jargon and place greater emphasis on capabilities, measures, and efficacy of the improvements across trend lines. One CISO based used the NIST cybersecurity framework as a model, which sets out Identify, Protect, Detect, Respond, and Recover stages.

That being said, most CISOs do not limit their communication options. They use various channels, including newsletters, emails, tabletop exercises, emergency call lists, and one-on-one lunches. Security roadshows can also become a vital component of overall cyber strategy as these event allow meaningful dialogs with business executives.

I had the privilege of having a one-on-one coffee with the government CISO of a country that holds incredible respect amongst other nations for their leadership and conduct. Due to the sensitivity of our conversation, its contents were off the record. However, I took away many gems from the seventy-five minutes we spent talking. The CISO shared that even he has not mastered the art of communicating with stakeholders. He is still learning every day, he said. His advice was to be a friend to them because it is all about building trust.

At a board level, the enterprise is paramount. At a personal level, the focus is on empathy and understanding.

A CISO's role is to be both a personal supporter and an enabler for the business. When the trust of the board is won, its members start to see the CISO as a trusted advisor. I will reveal more of my takeaways from this government CISO in Chapter Sixteen: 'Addressing the Conspiracy.'

The CTO of a multi-billion dollar company confided, "Although the board might not always understand, at least they listen. Try presenting to a team of engineers. They will just argue with everything you say. Every. Single. Thing."

When it comes to communicating with his board, this CTO follows a simple approach:

1. Keep it short. They have limited time.
2. Keep it simple. Ditch the jargon. Assume they have zero knowledge about the matter at hand and only mention as much as is required for an appreciation of the risk to be understood.
3. Keep it relevant. Cybersecurity doesn't seem relevant to some boards—but another company in the same industry with its name splashed across the headlines because it didn't follow proper due diligence is very relevant. A good communicator makes everything relatable.

In early 2019, I ran a Global CISOs Town Hall via Twitter with fifteen CISOs and influential cybersecurity leaders, which had 244,480 global impressions during the live event. One of the panelists was Darren Argyle, who now sits in the Group CISO office at a Global Bank as the Global Head Information and Cyber Security Officer. He proposed that when talking to boards, the CISOs use a blended communication strategy called SMS to keep communication with the board short and to the point.

1. Simplicity is all about being non-technical and losing the cyber jargon.
2. Measure by benchmarking against peer industry frameworks.
3. Stories are a good way to kick off communication.

Argyle said he used to overcomplicate things when engaging the board but eventually SMS worked for him.

Several CISOs shared successes in influencing people, executives, and board members without being dictatorial. There are certain statements that should not be made—for example, "This cannot be done."

In one of my earlier chats with Theo Nassiokas, he shared the importance of being a dynamic lateral thinker. "Unorthodox is the best word I can think of. You can't be a conventional thinker. You can't be a linear thinker," he said. "You've got to find ways around, over, under, or through things. The real solutions that will work best may not be the ones you think of immediately."

In a nutshell, CISO success is about coming up with solutions to complex problems that are also commercially feasible.

9

CHAPTER NINE

CISO and the Business: Power, Money, and Influence (Part II)

MORE MONEY, PLEASE

How do CISOs deal with the reality of budget constraints? There are various recommendations, a selection of which follow.

On planning ahead: It is important to have a clear, board-approved plan and strategy for the year and to get your team on board for it. Once that is done, match the budget to the plan and keep to it.

On managing budgets: CISOs advise managing the budget wisely. As companies go through massive digital transformation, they need to be prudent with costs. It is uncommon for CISOs to receive all the funding they want—or even half of it. It is important to understand priorities, rank them in order of criticality, look at the budget available, then fix the most important things first.

It is also important to respect vendors and to be honest with them. CISOs establish a good relationships and maintain a two-way discussion with vendors, and through this open dialog can seek discounts for future business and other beneficial outcomes.

Do not waste a vendor's time with countless demos and lunches. Open communication, especially on budget constraints and what needs to be achieved, will be more effective in determining what of benefit vendors can do.

On getting more budget: To achieve this, the key is to identify risk and business impact, as well as explore risk treatment options and costs. This information should be presented to the board, after which they can decide where the budget should be spent or whether they need to reconsider the CISO's budget.

Carmen Marsh, previously the chief operating officer (COO) at Cedaron Medical, a California-based company that creates cutting edge medical software solutions, shared that for most COOs, their goal is to achieve the highest earnings possible before taxes, depreciation, and amortization.

COOs need to show that they are running an effective and efficient operation for their company. This also means cutting budgets where appropriate, sometimes including IT projects, therefore impacting information security expense as well.

It is crucial for COOs to know what is non-negotiable. Every budget plan and business operation should factor in a company's security, privacy, confidentiality, integrity and availability of their data.

Elrich Engel, the head of cybersecurity at Vodafone, shared how he got creative with his commercial model.

The traditional model focuses on growing the asset base. Investing in new technologies or technical controls is often the easier task when it comes to building cyber capability. It was more straightforward securing capital expenditure (CapEx) budget.

As such, growing the functional personnel portfolio requires a different, and often more capitalised, resource-based approach as compared to organic growth. Over time Engel said he slowly wove key security processes into the existing project delivery framework. By "baking in" hard security gates into their standard delivery processes, he was able to slowly ensure that all the company's projects made provisions for security engagement in order to justify the resource requirements. This ensured that it adhere to secure-by-design principles and have mechanisms in place for timely identification of risks.

His department also recently moved to an all operating expenditure (OpEx) model for security personnel once they got finances were in place to consider security processes as business as usual (BAU).

He says the next phase is focused on moving towards more consumption or subscription-based models for service delivery. The primary focus would be on a total cost of ownership-based (TCO) approach, as opposed to CapEx vs. OpEx.

"You can't just keep asking for money," Nassiokas reminded, explaining that he has seen many CISOs go to organisations asking for anywhere between two to five hundred million dollars to fix security issues.

"Security problems are not fixed by large-scale deployments of technologies and technology transformations. Fixing security problems is done by chipping away the culture of the organisation to get the culture right. Next comes getting the processes right. Technology is the last in my view, not first," he said firmly.

If we haven't got our culture right, it indeed means that we don't have the right processes. This, in turn, means that the technology to automate those processes will not be appropriate either.

WIELDING INFLUENCE

I spoke to the former CISO of Israel Railways Israel Baron when he was in Singapore delivering a talk at the Singapore Rail Technology Conference. Baron is now with Cervello, an Israeli railway cybersecurity company that provides revolutionary and proven solutions to secure railways against cyberattacks.

Baron came from a background with the Israeli Ministry of Defense as the technology and information security officer and acted as the regulator of some of the biggest Israeli Defense Industries. He later took the first CISO position in Israel Railways and helped design a special security operations center (SOC), which is a first in their nation, and the world. This SOC monitors both the IT infrastructure and operational technology (OT), including Israel Railways signalling systems.

Currently, all of the world's railway systems monitor railway IT infrastructure. For years, IT networks and OT networks and systems have been running as independent networks. These environments are starting to converge.

For example, in the rail industry, critical systems, such as the signalling and other mission-critical systems, are connected to the IT environments. They get remote support from third-party suppliers, and sometimes even allow the integration of the IoT devices.

I was curious to find out how Baron managed to get such support, especially when his role was entirely new. I asked if he could walk me through exactly how he secured several tens of millions in funding..

"I took all of the company's core systems and worked together with one of the top security consulting companies in Israel for about half a year," he began. "We did a cyber survey that mapped all our weak points. In the end, not that it was a surprise, the conclusion was that cyber employees are needed from the construction of the department to staffing the SOC itself."

Baron said that on top of an SOC, there is a need to monitor operational systems as well. He went to the CEO and board to present the conclusions.

"But how did you do that?" I wanted details.

Baron elaborated, "I explained to the board that we have to look for solutions. Not for now, but for 20 years from now. I showed them what we have is a strategic infrastructure. It is not just about buying any intellectual properties. My role was not to secure the safety of systems, but rather to secure the safety of people."

He continued, "If an attacker hits the SAP database, it doesn't matter. But if this person hits the signalling system, which controls the trains, this is the most critical infrastructure that can result in the loss of human lives."

Barron's passion was evident as he stressed that when it comes to critical infrastructure, there were two things to consider:

1. Do everything possible to prevent the loss of priceless human life.
2. Prevent operational and reputational damage to the company should there be an attack on the train systems.

His was a challenge to the board, demanding of its members whether they had done the best they could to prevent such damages.

Baron detailed for the board the impact of the exact threats that the members would be responsible for. "I described what can happen when I got into the signalling system. I described how I could easily cause a train accident by moving the switch in front of a moving passenger train, derailing it whilst 1,000 or perhaps even 2,000 are people on board."

After painting the different scenarios in a realistic and graphic way to drive home his point, Baron also highlighted the importance of taking a holistic point of view and giving management a solution that addresses both the IT and OT environments.

#Understanding the Board

Now for the non-critical infrastructure environments—how would one articulate this to the board? Well, this is dependent on the types of people who sit on this board, their formal or informal styles, personalities, and preferred ways of communication, as well as their backgrounds. Are they IT focused or are they more oriented toward operations?

Everyone has a different view and opinion.

Chin Kiat, the CISO of DHL Express, told me that first he needs to find out whom within the board is the opinion leader with the most influencing power. Every committee has someone who is the most vocal, who asks lots of questions, and who will present challenges. There also be those who just go along with the rest.

Once he has identified the thought leader, he establishes a relationship with him or her. It has to become an effective relationship, so it is important to take the time to understand whether this person is technical and whether he should focus on statistics or on value.

It is important for him to link the cybersecurity problem to the business pain point, explaining how investing in security posture improvement will also help business success, why the organisation should invest in cybersecurity, and why they should do so there and then.

He said he expects to hear questions such as "Why do it now? We've invested in cybersecurity in the last ten years. What will the return on investment (RoI) be?" When he does, he says he turns it into a different dialogue. These days, one can talk about the typical risk heat maps, focusing on consequences and likelihood, he said. There's even the Factor Analysis of Information Risk (FAIR) methodology or other statistical models. But for Chin Kiat, the focus is on return of value (RoV).

"When we talk about budget in a defense department in the government, we don't talk about RoI. Why? It is the duty of the government to defend the country when there's a war." It is definitely a value for the government to secure the nation.

"Similarly, right now we are in cyber war. Do you think

the police or the government is going to defend you when you're under cyberattack? You must defend yourself; you need a defence budget," Chin Kiat explained.

"How much budget? That's up to you to decide based on the organisation's complexity and risk appetite," he said. "But it is an obligation to have a defence budget, called the cybersecurity budget, so that you are able to protect and defend the organisation in the event of an attack. If you don't have that capability, that is the risk to the business. How are you going to measure the risk? Well, there are lots of examples in the news."

It's true. Most people already know about risk and cybersecurity because of the media. The Cisco 2018 Asia-Pacific Security Capabilities Benchmark Study, which compared their cybersecurity standing of 11 countries, revealed that Australia is the nation most under attack, with 90 per cent of Australian companies reporting that they receive up to 5,000 threats a day. Despite this, the main question remains, do these organisations understand the relevance of it? Is there any inside information to evaluate the stage of risk at which these companies sit?

It is easy to reference other companies that have been breached. Yet, it is not personal enough or hit close enough to home. Chin Kiat believes in using internal data. "I will look at the organisation's data to tell the story, from the number of systems that are vulnerable to the number of systems attacked internally or externally, and the number of threats we are facing," he said.

"Basically, they can't simply say 'No' without applying their due diligence as the management. When a breach happens, they are accountable for their decision. I'm not referencing external data that is not relevant to the organisation, but data from our internal organisation. If they don't make the right decision, and an incident happens, an authority may want to investigate," he states.

What Chin Kiat said next added to what Nassiokas said earlier about plausible deniability. "As a board, it has an obligation to apply due diligence. And now that I've made it transparent to them, this could potentially become a problem if they have accepted the risk. Hence, it is also important that they justify their decision. If not this becomes challenging."

For any identified problems, it is important to have internal data documented—not just external data. It can be tricky for most organisations to achieve that transparency, but without it, organisations cannot move past the first step.

The good thing about transparency is that once it exists, it breeds responsibility. The problem has been identified, now it must be fixed.

#Networking

This is where networking comes in. It is important for the to pinpoint where contributions and influence will make the biggest impact.

Phil Agcaoili has been the CISO of four companies and chairman of the Ponemon Institute Fellows, based in Atlanta, Georgia. He emphasised to me that the biggest advice he could impart to other CISOs was, "Be likable!"

Chin Kiat walked me through how CISOs can expand their circle of influence. They should work toward expanding their circles of influence to include ten new people. These ten together have a hundred people in their spheres of influence who can be drawn in. The more people in a circle, the more supportive and trusting of a CISO's message they will be.

The former CISO of the State of Michigan puts it this way, "If you can't beat them, join them."

Yes, it might be all too familiar, hearing CISOs complain that they don't have anywhere near the funding or staff required to protect systems. The solution is to get on the boat leaving the dock! Leverage "hot button" issues that are currently receiving funding.

CISOs can make sure that security is built into these funded projects. They can gain a seat at the table as a committee member or key resource for important initiatives. CISOs can go beyond basic duties and help the wider technology and business teams succeed. This may include basic technology maintenance projects. Many security breaches are actually the result of simple neglect or of technical professionals who did not do their job in areas such as patching server vulnerabilities.

Security leaders can help technology leaders be more successful, thus better protecting the enterprise at the same time. "The money doesn't need to be in your budget. You may provide better cyber defense by helping others be more successful," Dan Lohrmann concluded.

10

CHAPTER TEN

CISO and the Business: The Art of Storytelling

In the earlier chapters, whilst we looked at the various ways CISOs manage and influence upwards, in this chapter, we focus on doing something no CISO should never neglect: constantly telling a story.

"Stories are about people. Describe them so we can see them," said Patricia Fripp, the first female president of the National Speakers Association.

Now, I understand why some people in the security industry don't like stories. They believe they trivialise what is already not easy to understand. My take is use whatever means work with the audience. A story can illustrate what could happen and the impact that could have on an organisation.

CISOs need to remind businesses of what they have accomplished, including the risks they have reduced and the benefits they have brought to their organisations, as well as presenting numbers such as how many attacks firewalls have blocked or the number of viruses cured by antivirus programs

Chin Kiat explained how he does this. DHL has a number of customer-facing websites that process global online shipping orders. In the logistics industry, November and December's festive season are peak periods. It is particularly important for DHL to deliver not just on the high volume of shipments and revenue, but to also put smiles on people's faces at the arrival of millions of holiday parcels, packages, and letters.

With the right numbers, Chin Kiat was able to tell his management that in a peak period, with the right defence tool, his team was able to ascertain and mitigate a number of cyber-threats that came from specific countries who were targeting them.

As a result, there was no disruption to shipments and DHL was able to complete the peak period's high volume of shipments, earning millions of dollars in revenue. This was the beneficial outcome of security investment—enhanced resilience operations and a profitable network. However, if any of the attacks succeeded, the impact on revenue would have been catastrophic.

This story is told in the language his business understands and will remember, especially when qualified with exact numbers.

Chin Kiat says when the management sees a consistent demonstration of this and trust is established, they know it will be of true importance when he asks for something. They understand that investing in security will either contribute directly or indirectly to revenue production, or to secure the revenue. They can trust to see the same quality results and be sure they will get consistent quality for the approved funding.

SAGE ADVICE

#1

Cloud services are always working and available anywhere—that's part of their value proposition. But when a user logs into a cloud service, it is extremely difficult to verify his or her identity beyond an old-fashioned username and password.

We used to talk about "defence in depth." The concept was simple. We buried our important data deep within the network. This usually meant racks of servers sitting in data centres that were either hosted on-site or at a co-location. We put layers of controls around these servers, such as firewalls, access control lists, and multiple tiers to keep the data on the servers safe. The idea was to make it as hard as possible for unauthorised persons to get anywhere near the sensitive data.

Things that once were physical in nature are now gossamer-thin. One could see the servers and touch the racks. Entry into the data centre was required to make any alterations. Today, with cloud everywhere, we can no longer put physical walls or security layers around our prized assets. All you need is a web browser and login credentials and you're in.

#2

Elon Musk is a brilliant man who has founded cutting-edge businesses such as PayPal and Tesla. There are not many automotive innovators like Musk out there, right? Even so, would his company Tesla ever built cars without brakes? Seatbelts? Airbags?

These necessary safety measures mimic proactive security in business settings. Security is not slowing the business down, but speeding up growth because protective mechanisms are in place. If you run a business without properly integrating proactive security, it would be the same as a Tesla on the road without brakes or seatbelts.

If you want to build a business that will never get off the ground, build it without security.

#3

Having good security is like having a torch as you walk through a dark area. It won't always protect you, but it will help show many of the dangers, so you can act wisely.

#4

At a street crossing, will you wait if the light shows the red man and only go when it displays the green man? The question is how much risk are you willing to take?

#5

A very low-profile phishing attack occurred targeting SWIFT analysts who dealt with large global transactions. These SWIFT analysts were not thinking about such an attack and were not disturbed by criminals who went low and slow for about 18 to 24 months.

Transactions would be flagged if the attack topped a certain amount. The attackers stayed beneath that threshold and so moved hundreds of millions before they were caught.

This illustrates the importance of educating and training your employees on the potential dangers surrounding them. They need to know the enemy and understand why and how they could be attacked.

"Most people know what happened with WannaCry and NotPetya because the monetary losses of all companies were widely reported. The entanglement of our interconnected world brought everything to a crashing halt in just a few hours.

Global events such as the WannaCry cyberattack incurred losses of over $120 million pounds. NotPetya was the most destructive and costly, causing more than eight billion dollars in damage worldwide.

The IoT is not governed by appropriate regulations, it's unknown where the supply chain starts or ends, and there are no standards around the level of security that must be assigned to the IoT's "things." But until something occurs, most people think, "Well, that's never going to happen to me."

People understand financial impact and dollars. It is the language of the CFO. By showing what a certain level of risk means in the enterprise risk framework, CISOs can be specific and use statistics—for example, "There's a 20 per cent chance we can lose five hundred million dollars in the case of a data breach under current controls." CISOs can use also use heat maps that project consequences and likelihoods—for example, "This sits in the amber state of risk for my organisation."

There are different schools of thought for risks but at the end of the day, it's about CISOs working within the maturity and appetite of their organisations and ensuring their CFOs, CROs, or CEOs understand the impact to the enterprise risk framework and what this ultimately means to the business.

11

CHAPTER ELEVEN

The Daily CISO: BAU Survival (Part I)

In the past few chapters, we have looked at different ways of communicating upward. But in addition, a CISO needs to collaborate.

In one of my conversations with Theo Nassiokas, he noted that whilst there are CISOs who are great at managing upward, thereby accruing a lot of political influence, they are disliked by their own teams because they use threats and intimidation to manage outcomes.

There are also CISOs who are great at managing downward and their teams love them. However, they are terrible at managing upward and, therefore, don't have political influence.

"The reality is that you need both. You need to be able to manage downwards, sideways, and upwards," says Nassiokas. "You need to have the support and trust of your team, your stakeholders and the board. You need all of that. Nothing will fool you faster or bring you undone than not knowing your true nature."

Nassiokas then spoke of Sun Tzu, the revered Chinese general and military figure, who wrote influential military tome The Art of War. He repeated one of Sun Tzu's quotes really struck him: "If you know the enemy and know yourself, you need not fear the result of a hundred battles."

In the security context, we don't really have any enemies in the private sector. However, we need to know our stakeholders. We need to know our business and understand what commercial feasibility means in the context of our organisation.

Nassiokas explained, "Being a good CISO means you must really understand and accept your strengths and weaknesses. If you understand your strengths, you would know how to use them to successfully accomplish your role. More importantly, when you understand your weaknesses, you know who to hire people who are stronger in those areas to help you, then support and trust them. If you succeed, you all succeed. If you fail, you all fail. You don't throw your team under a bus. You stand up and be a leader."

He continued gravely, "Nobody can be good at everything. That's why when you've got a great team comprised of the right people, with right skillsets and within the right culture, you have a situation where the whole is truly greater than the sum of the parts."

With that, let's take off the filters and examine the different ways security leaders can miss the mark.

GETTING REAL

The question, "Where have our Security Leaders been getting it wrong?" leads to a multitude of reasons. Tony Vizza, APAC director for (ISC)², an international non-profit membership association best known for the acclaimed Certified Information Systems Security Professional certification, spotlighted a number of them for me:

- By expecting an organisation's executive to understand, or even care about cybersecurity beyond doing the barest minimum necessary to comply with regulatory requirements—and even there, often missing the mark.
- By not relating cyber risks in terms and constructs that an executive considers on par with other organisational risk such as fire, theft, natural disasters, or employee injury. Because of this failing, decision makers see cyber risk as something "that won't happen to us."
- By relying on so-called solutions provided by sales-savvy vendors who have fantastic sales teams and compelling arguments for why a CISO should hand over a sum of money for "peace of mind." These solutions are often point products to solve a specific need and little consideration is placed on how that solution will help achieve an organisation-wide reduction of cyber risks.
- By expecting IT staff to understand cybersecurity concepts. IT and cybersecurity engineers work in different disciplines. A system administrator is not a cybersecurity engineer any more than an automotive mechanic is an aviation technician. Sure, there are areas of broad overlap, but the nuances of IT and InfoSec are vastly different—for example, both cars and planes have

engines but the engines themselves are not alike.

- By depending on cybersecurity staff who lack the necessary skills, experience, and certification.
- By treating employees of the organisation who have committed an act of poor cybersecurity hygiene with contempt and derision.
- By not investing in cyber insurance, or not purchasing enough of it. Some policies pay out their limit before initial investigations are even completed.

In subsequent chapters, we will address ways in which CISOs have bridged some of these gaps successfully.

#Managing Sideways

I met with the CISO of one of the world's most renowned airlines in a mini alley of cafes that bustled with people meet whilst having breakfast or tea.

Safety is this airline's number one priority, so, I won't be revealing his name or the name of the company. This restriction, however, allows the authenticity of conversation to be retained.

Despite all his achievements, this CISO was down to earth. He made me feel immediately comfortable. Both of us ordered a green tea with a pot of honey and he went right into it.

During all my previous chats with CISOs, we focused on their schools of thought. In this conversation, however, he turned the tables and asked me what was missing in many current cyber functions. He went on to summarise the following three points.

First, there is a difference between the traditional cybersecurity function that has a good understanding of the business versus an influential business function that solves the business security problem across the company. It is a subtle difference, but important and mindset altering. It is also especially crucial when trying to drive a massive change within an organisation.

Second, 70 per cent of traditional cyber strategy is the same across all companies. Traditional cyber thinking is all about managing risk, employing technology, the awareness level of the business's employees, industry influence, third parties, and more.

The challenge that many companies face is adopting the right strategy to be successful, and the top layer of this is strategy execution.

My ears perked up, as it was the same secret that Sander Swinkels shared with me. I was curious what execution would look like, because the bigger the company, the more complexity and challenge in the implementation.

Third, the key element of execution is the ability to influence.

How do you set up a business function that is influential? In the digital age, cyber directly impacts every employee. The CISO needs to influence every employee, he stressed. More brainpower needs to be put into figuring out how to do so.

For example, if a CEO has seven direct reports, how many of them are advocates? If you are lucky, one of the seven will drive your topic, but as part of driving yours, he will also drive others. His peers will drive their own topics. There will be trade-offs and the CISO will not get 100 per cent influence.

The key element is that at the CEO's table, you need to

have as many advocates as possible to sustain influence.

As if reading my mind, this CISO continued, "So how do you get that? That's the magic. Every organisation has different ingredients. There are things that you can change and there are things that you can't change. There's no silver bullet in raising top-down support. It is much easier said than done."

He said, "Imagine this: You need certain sort of soil to grow certain plants. Since you don't have the soil, you change the plants you grow. You need to understand the environment you are growing in so that you can nurture the right things. When I drive cyber change or initiative, I always find a business executive sponsor. I never do it alone."

At an international digital company, he drove the personal information (PI) improvements in relation to privacy and data protection. He didn't just do it alone or in tandem with Legal. He understood that PI has reputational consequences, so he saw it is a marketing project. He formed a bond with the chief marketing officer and they worked together. To him, it was about finding the right angle, the synergy, and right trigger for the appetite.

He went on to talk about the importance of assessing the three different levels: the unsupportive, the neutral, and the supportive.

The CISO posed this question to me: "How many hours a day do you put in thinking about different things and trying to build synergy?"

This CISO never took the Payment Card Industry Data Security Standard (PCI DSS) on. In fact, he said he spends more time figuring out how not take things on as a way to influence synergistic partnerships. For example, with PCI-DSS, he always involves the CFO and brings the discussion directly to him. After explaining the challenge, the CISO asks the CFO to nominate a leader and sponsor, thus making it finance project.

As you can see by these two examples, he managed to shift two CxOs to go from initially being unsupportive to perhaps harbouring some remaining uncertainty, to being his advocate.

"You have to find a compelling enough reason that they must listen to," he explained. It wasn't due to how much he pushed. Successful execution always goes beyond the individual. It is more than just actively trying to take things on. As a result of the CISO having such an approach, it eventually led to human resources (HR) approaching him to roll out his idea of making cybersecurity a KPI for every employee. That is a milestone and an invaluable achievement within such a large organisation.

The CISO's idea was cleverly achieved by embedding the cybersecurity message and measures the same way other key behaviours are measured. In this way, there is no need to train staff or managers. Expected behaviour guidelines are set forth. To assess security-aware behaviour, only one measure is needed: "Have you done it or not done it?"

I asked for an example. He said there are KPIs for various employee layers. The business executives, for instance, are two levels down from the group CEO. The level of KPI allocates the responsibility of controlling their own people, suppliers, and systems back to them. Their own business suppliers can share a certain set of information for them to action certain things for customers. The system supports the process for defined business functions. As the business owner, they will be able to influence their respective KPI, and that is the measure.

Even as the C-level grows in power, it doesn't necessarily

mean that CISOs will be successful. To illustrate this point, this CISO mentioned surfing. A good wave is important. However, as the waves approach, a surfer needs to know how to paddle hard, time it right, then stand up with the correct foot positioning and posture. Once all of that is achieved when the wave arrives, the surfer will be able to have the desired outcome, which is to enjoy the surf.

Likewise, if the CISO doesn't get things right, or do these things well enough—i.e. he is not able to educate leadership or get the execution right—when the sea level rises, the CISO will be swallowed by the wave.

12

CHAPTER TWELVE

The Daily CISO: BAU Survival (Part II)

#Let the Vendor Speak

One of the challenges mentioned in the previous chapter was CISOs' need for vendors to consider the fit of their solution with the overall organisation's objective in reducing cyber risks. I thought it would be interesting to hear some honest opinions from a few of our vendors on the security challenges they currently see.

A CEO of a renowned security vendor based in Hong Kong said he felt his biggest challenge was the absolute blurriness of the perimeter, coupled with heterogeneous work environments.

He provided an example.

Some departments prefer MAC for obvious reasons, others prefer Linux and most are a Windows environment on the client end. Add to it all, bringing your own devices (BYOD) makes it extremely, if not completely impossible, to ensure IP protection and compliance whilst still managing flexible expectations for a Millennial workforce.

There seems to be no perfect solution at this stage, he said. Even the CEO is not satisfied with current strategies that include layers upon layers of technology and checks. He questioned its efficiency, especially when things evolve so quickly due to the large number of ventures and their sub-cultures.

When quizzed on the biggest problem a CTO faces on the vendor end, he said it is getting around the "you're trying to sell me something" assumption. We can all understand why the walls go up. At the end of the day, if CISOs are not listening to what a vendor offers, then why meet with them at all? He shared that he has met with government agencies that told him, "We're safe because we use an air gap." Whilst an air gap might help, it does not provide complete security. Most likely, it will merely change the way bad actors will plan their attack.

Another issue is caused by pricing,, which is often treated as a box ticking exercise. Got VPN? Check. Got firewall? Check. However, these components are not all the same. Their capabilities differ dramatically. Many vendors have won deals by offering 80 per cent of the security of better products at 50 per cent of the price. This might sound like a good deal, until you realise that a 20 per cent gap in security is actually huge.

To be honest, this issue goes both ways. The customers always think they are right, but then, so do the vendors.

At the end of the day, it is important for both the CISO

and vendor to establish a trusted relationship in which there is consistency and open communication. If your role involves close interaction with cyber risk leaders, the Bonus Chapter Selling to the CISO will reveal the truth about what CISOs look for in their partners.

#The Resource Problem

One of the top struggles we deal with in our industry is people. Most executives who I've spoken to would agree that getting the right human resources is challenging these days.

Hiring a good and experienced person who understands technology is hard. Getting the right culture fit for a team is even tougher. Corey Mahan, information security manager at Tyro Payments, Australia's business-only bank, agreed that recruitment can be challenging when you're looking for more than just the cultural status quo. "We don't just look for a culture-fit. We look for culture-add!" he said.

Most cyber risk leaders' advise hiring managers to form a clear idea of whom they need and to be open with candidates, encouraging them to pose as many questions as they are asked.

One of the CISOs said a common question he usually asks is, "What do you wish to be a typical day?" The reply, he told me provides excellent insight into the candidate's experiences and knowledge.

In one of my chats with another CISO, he shared a personal, favourite interview question that he usually poses to his candidates: "How do you deal with frustrations?" As an insider, I know it is common for security professionals to experience frustration. I've felt it after proposing something that my company needs for protection, only to hear "No" without any justification. There are those absorb frustrations, until they come out in negative ways.

Hiring managers must assess candidates on technical skills, as well as from a psychological perspective. They must also root out those who are fixated on collecting certifications but have only theoretical, rather than real-life, hands-on experience. The ability to understand technical environments is more crucial.

I previously worked as cybersecurity recruitment specialist, and I can attest that it took real passion for the industry and lots of diligent hard work to carve a niche for myself when awareness of cybersecurity was still quite new.

Cybersecurity recruiters must truly know the ins and outs of the cyber talent market. They must, understand the motivations of individuals, know how to give professionals proper advice on career decisions, and provide insight on the relevant cyber skills required. In this way, they act as advisors or business partners to CISOs, heads of security, or other hiring managers.

No matter how our world advances, we can never take away the human factor. People matter. People will remain our greatest resource. There are several things to be mindful of when identifying, attracting, and retaining talent.

FINDING THE RIGHT PEOPLE

Let's face the facts: Highly skilled people are always in demand and it is a competitive job market. As hiring managers, we must look at our current environment to determine how conducive it is to nurturing productive talent.

I know some intelligent minds from the military who have told me that when it comes to transitioning and trying to thrive in the corporate world, the struggle is real.

In the military, team collaboration, teamwork, and team interest are at the forefront. There is usually a deep trust between co-workers. Unfortunately, most said they had to let go of these expectations and behaviours in the commercial sector. The focus on individual key performance indicators, backstabbing, and office politics do bring out the ugliness in human nature. One executive told me, "You do not want to be eaten!"

What makes a company attractive to work for? A CEO in Singapore spoke to me about when he took a role with a global company as its first hire in Asia and was tasked with building the market up from scratch.

Everything happened fast in this global start-up. Whilst previously holding the position of General Manager (GM), he hadn't been trained yet and, unlike the rest of the global team who were engineers, he wasn't technical.On one of his first dealings with a potential client, he was required to fill in a certain document, which he did to the best of his ability. Unable to complete the rest of the more technical elements, he sent an email to the whole organisation, explaining that he was new and unsure about a number of the requirements he had to submit the next day. "Is there anyone who is willing to help?" the GM asked.

By then, it was 5 a.m. He told me that he went to sleep for an hour and a half. When he awoke, to his astonishment, his inbox was completely full. His colleagues from the other side of the world responded to his call for help even though they didn't know him. They felt an obligation to step up and help a fellow co-worker.

To this day, the culture in that company, where they never let a colleague down, has left a strong impression on him. He has never forgotten the feeling of support.

As part of my role, I have visited many workplaces and

observed the different environments that people work in. Some of the memorable ones included people whizzing past on electric scooters whilst using their Bluetooth headsets on a call. In one company, it wasn't uncommon to see colleagues playing table tennis or trying to score more goals in the office mini-golf putt whilst in the middle of business calls.

Another CEO shared that in one company in Washington, D.C. - a place renowned for innovators - at lunchtime, they could be seen coming out of meeting rooms to duel with Star Wars lightsabers. The same company also had bulldog that sometimes would be seen on a skateboard.

This company won hands-down against competitors when it came to rapid growth and also succeeded in securing like-minded talent through word of mouth. As conventional wisdom says, "Good people know good people." The employees genuinely loved both what they did and the company they worked for.

On top of attracting the right talent, CISOs also must aim retain it.

SAFEGUARD THE CULTURE

Sander Swinkels told me that at the global innovative giant Palantir, he was charged with building out the entire APAC market. No one was hired unless there was a final interview with one of the founders to ensure there was a cultural fit.

One of the benefits of a good organisational culture is the ability to attain more effective results in improving the cybersecurity capability.

#Building the Right Culture

Head of Information Security (APAC) of a financial services firm in Australia, Laura Lees, shared her observations. "It is like the carrot versus the stick. A fitting example where I have seen an encouraging change in responsiveness is where people are consistently rewarded through positive reinforcement. For example, developers are encouraged to write secure code through gamification in which tournaments are run and prizes awarded. Developers get really engaged; they want to learn and get better at security, so they can win the best prizes."

Lees elaborated that the converse is true when people are reprimanded for negative behaviours that increase the risk of compromise. She gave an example in which development phases might be delayed when developers have not been able to demonstrate secure code. Costs increase due to repeated testing and rewriting. As a result, people become reluctant to share when they have faced an issue or when a threat has been realised, for fear of getting into trouble. Unfortunately, this type of behaviour increases risks as the organisation loses visibility of the true threats.

People are integral to improving the cyber capability of any organisation. Laura added that personally, when she feels valued and is supported by management to further develop her skills, she is highly motivated to deliver better outcomes. She is more engaged and feels a stronger connection to the organisation's strategic goals.

Lees shared that she has worked with managers who are genuinely encouraging the furtherance of her career, helping develop her for her next role, regardless of whether or not it is an internal one.

The mark of great leadership is where their people supersede even the leaders themselves.

#Invest in Your People

Cybercriminals can industrialise and exploit gaps in the cyber industry because many people lack the awareness and pragmatic knowledge to detect, stop, and defend. Therefore, it is important to equip our talent pool with the right and updated skills.

One of the CISOs I spoke with attributed his current success to a previous company that gave him many internal opportunities. Whilst this is ideal, unfortunately, it is more tempting to companies to go after external, ready-made products. CISOs and security professionals are routinely headhunted for the same roles in different companies.

However, for the betterment of our entire industry, and especially in the cybersecurity space, it is more beneficial to have a plethora of diverse backgrounds and perspectives to tackle the same problem. Sustainability required us to invest keeping staffers' skill sets current and grooming them for leadership roles. This will grow the cyber talent pool whilst bridging current gaps in the cyber industry.

What's next after CISOs have attracted and invested in the right talent? We want to keep those hires fulfilled for as long as possible. A good retention tip is to establish real connections with people.

VITALITY OF CONNECTION

When a leader is able to make people work together rather than against each other, it is a marvellous accomplishment that enables them to position cybersecurity requirements at the beginning of every stage. In the long run, this saves the organisation both time and money.

A CISO from Israel told me he sits with his team to understand their needs and figure out their problems. He learnt this working in the defense industry, developing satellites, unmanned vehicles, missiles, and more. Before he became a CISO, he asked engineers to walk him through their projects, and gained invaluable insights and new perspectives from their answers, he said. It was through building such connections that he concluded this was the best way to harness people together to get the job done.

CHAPTER THIRTEEN

13

The Daily CISO: BAU Survival (Part III)

IT'S AUDIT TIME

Dealing with inexperienced auditors can be a challenge for CISOs. There have been many occasions when audits have been assigned to junior auditors with little or no practical information security experience, which resulted in the reporting of incorrect findings.

One CISO gave an example: Controls are implemented to reduce risk and the auditor assesses and reports on the effectiveness of the control. On numerous occasions, he stated, the auditors cited an absence of a control and stated that risk was caused therefore.

"I don't believe that is correct, and this can cause organisations to focus on implementing point solutions rather than identifying, assessing, and managing the risks," the CISO explained.

One of the CISOs, who was formerly an auditor, shared his take on the best way to pass an audit, which is to run it as a project with budget and resources. That, however, is a short-term win, he told me. The best way is to perceive audits as BAU and do what the standard in question requires. If the CISO inculcates a culture of "This is just what we do," an audit can happen anytime without concern or issues.

Hai Tran, the CISO from the Western Australia Police, believes that audits are fantastic opportunities to improve security. They can also provide the chance to communicate with senior management about risk and other issues.

So, yes, audits can be an effective way of uncovering control gaps. In fact, they can be leveraged to identify gaps that need extra security resources. One government CISO even likened it to "using lemons to create lemonade" that can help fund meaningful projects.

All of this, however, is highly dependent on the strength of the audit team. An ideal audit team is one that establishes a collaborative relationship with a common risk language.

"No one wants to work with the audit team that runs on the 'gotcha!' operating model. It doesn't help anyone," notes Theo Nassiokas.

I had a productive discussion with the global CISO of a German company comprised of several hundred thousand employees who wanted to remain anonymous. He shared these thoughts on auditors, "They are the easiest people to work with. They audit according to a standard and it depends on my objective. I can decide to get a good audit result if I want to."

He walked me through the steps. Before an audit, he receives the audit notification and information on the scope of the audit and the evidence they want to see. This helps him make the necessary preparations to ensure a positive result. However, this just goes to show that the outcome of the audit can be controlled.

This CISO gave his honest opinion, "I don't really appreciate how we do audits with a lot of the auditing firms these days. Many times, they tell me the problem that I've already known for a long time. There's just no point in repeating the issue, as it will not add any value to how you can further enhance the cybersecurity or compliance posture of the company."

Auditors are hired to assess the risk of the company. They are educated on risk and compliance and mainly focus on controls. They do not know what cyber threats or cyber attacks look like.

He believes that it is important for the auditors to upskill themselves to see the real issues beyond the checklist. If they are able to put on the cybersecurity lens to ascertain beyond compliance, including the risk and threats, as well as how effectively people can misuse these gaps, it will result in a more realistic impression of the overall security status, thereby presenting a lot more value to senior leadership.

"Don't just say, 'I don't see good user access control.' Tell me, perhaps, that with testing and using this tool to exploit, you are able to elevate yourself to an administrator. Now, sometimes you might not be able to do that, but at least, give pragmatic advice. It gives the findings more meaning. Don't just state evidence," he said.

In fact, maybe it's time to improve the audit standard to

embed a more pragmatic way of assessing the cyber risk, instead of merely assessing it on ISO 27001, COBIT, or other traditional standards.

This will also help CISOs obtain the right attention from senior leadership, making it easier to convince them to support efforts to fix problems.

YOU DID WHAT?

In the last few chapters, we have seen CISOs take apart some of the key challenges of the cybersecurity industry. A CISO needs to be prepared to take risks to do what's best for the company, not what's best for themselves. You need gumption and bravery to do this.

I chatted with a CEO who runs a global consulting firm, and she shared a real story of how one of her clients was phished twice in one year—once for $70,000 USD, then for more than $200,000 USD!

The first attempt was stopped before the money was transferred to an offshore account. Unfortunately, $200,000 USD was wired out by the CFO, who had been with the company for nearly 20 years.

The CEO was going to fire the CFO and put all the blame on him. What was even more incredible was the CEO's statement, "Well, this is just part of running the business!"

Fortunately, since then, they have taken on a cyber risk program; they are now implementing critical security safeguards that include investing in companywide security training awareness.

This reminded me of when I received a panicked call from a company that was phished, resulting in a lot of money being sent to the phishers. I suggested that an investigation be conducted to isolate where and how the attacker got in, as well as identify any gaps in their processes, technology, or people.

A day passed, then another. I tried calling but no one picked up. When I finally got through, they told me they didn't need my help any longer. They had rebooted their entire IT system, wiping off all the evidence, and bought a whole lot of new and expensive firewalls.

My recommendation would have been to first validate that theirs was an actual incident. I would next look at ways to ensure the incident was contained, then investigate the system that had been compromised. Once the company traced it back to the account to understand how it happened, it could add a validation and verification control in the payment authorisation process to ensure phishing didn't happen again. Sadly, throwing money at new technology does not solve the root problem.

We have heard of phishing attempts as part of security awareness programs. Some individuals targeted were at senior levels in these organisations. One CTO told me that a vice president was sent a phishing email. He promptly opened it but couldn't get the attachment to load, so he tried over and over again. Finally, in frustration, he forwarded it to his PA to see whether she could open it.

It's good that with media attention, more companies now understand the importance of security awareness training for their staffs. However, there is still a lack of full commitment to give it to them. The aforementioned company arranged to have a few of the senior leaders attend a single training session and hoped that whatever the leaders learnt would cascade down to the staff, but it never happened.

One C-suite executive shared a story about an APAC

company that agreed to build a whole security capability. They budgeted for a SOC, an operating model, and security strategy that included the architecture, requirements, programs, and projects. During this build-up, more than 20 people were hired. Whilst coming up with the strategy and its early implementation, the company spent close to a million dollars.

Shortly thereafter, the company leaders decided they didn't like the plan, threw it aside, fired the new hires, and went back to treating security as a tactical function.

The executive explained that, unfortunately, this was a classic example of leadership paying lip service to security--saying "Yes, we think security is strategic," when in fact, they did not. They let the people heading the security capability run with it until, at some point, they decided to divert the finances elsewhere.

There have been other times, he said, that he saw companies spend six to seven figures on new security capabilities only to shortly throw it all away. Perhaps, if the strategic impact was communicated properly to the board at the beginning and the success of the security program was directly tied to that of the board, then these programs would be sustained and would thrive.

14

CHAPTER FOURTEEN

Global Perspectives: Staying Ahead (Part I)

THE CISOS' DILEMMA

In the early days of IT security, people lived in blissful oblivion. We thought that with a basic virus scan of our laptops and by not clicking suspicious links, we would be safe. As cyberattacks and IT disruptions increased, eyes were opened to reality, and as a result, fear and uncertainty was perpetuated amongst the general public.

Now, citizens are more willing and likely to purchase security products. Yet, most people do not know what they need or what they should buy. To feel protected, they choose products impulsively and real protection is not achieved.

People must identify which aspects of their information they should protect to safeguard themselves. Is it your financial details, your private information, or your children's photos?

A CEO based in Europe reminded me that in some countries, if you are wealthy, it's not good for people to know what your children look like or which school they attend. Likewise, for those who do sensitive work in the security domain, it is also important to protect our families and friends. There are those who will be unhappy about the types of problems we solve, such as finding hackers or identifying crimes. For this reason, there are many who stay off Facebook or other forms of social media. As I wrote above, one needs to identify what one wants to protect.

Sander Swinkels, former GM of Palantir Technologies, was the company's first SEA hire. Palantir specialises in big data analytics and is known as one of the most secretive companies in Silicon Valley.

Palantir is partially funded by InQtel, the venture arm of the U.S. Central Intelligence Agency. The company works with intelligence agencies, the military, and law enforcement such as the Los Angeles and New York police departments, and countless other crime fighters. It also serves giants such as Airbus and international banks. Additionally, media sources have claimed that Palantir was used to track down Osama Bin Laden.

Understandably, Swinkels did not confirm any of Palantir's work. As the previous GM in charge of Southeast Asian growth, he shared the work they did made an impact that he is proud of. He says it is crucial for service providers to employ solutions that work. However, despite an increasing number of cyber solutions, the people buying them may remain unsafe.

#The Evolution of Privacy

Privacy is not a new topic, especially for Europeans with memories of previous state-controlled governments and regimes. The monitoring of people and privacy is especially sensitive. But Swinkels compares the changing concept of privacy to a community of small villages in Europe seventy years ago.

At that time, privacy was no issue at all. Everyone knew everything about each other, their neighbours, their friends, their relations, and more. Nobody cared about privacy, or the lack thereof. In fact, information sharing provided a safety net for people during times of difficulty.

However, with nationalist regimes, World War II, and rise of the German Ministry of State Security and the Nazi Secret Police, the concepts of privacy and identity changed as data was collected and used to monitor the citizenry.

"One thing that has always intrigued me, is how much people are willing to share about other people," Swinkels reflected. "All the systems, although government initiated, worked not because of force, but because of the willingness of people to report and share things. It's in the human nature to share about what others are doing and to talk about what other people are doing."

In the present day, with social media, everyone is even more willing to share what's going on in their personal lives. If you were to talk to the younger generations, they don't care whether their posts will stay online for eternity.

The current dilemma is caused by a deep mistrust of government on what it can monitor. Businesses and organizations are now rethinking how they approach information security because of the European General Data Protection Regulations (GDPR), which became enforced in 2018. The object of the GDPR is to equalize European privacy laws, thus ensuring the privacy of EU citizen's personal information.

"One thing I know is that most governments have some form of checks and balances in place, and regulations on how data can be stored, how it can be used, and for what purpose. Depending on the regime, there's also some form of a supervisory board, or control mechanism to test, to a certain extent, if there's any misuse of the data," Swinkels stated. "Even in countries where people are repressed, there still is some kind of regulation and if misused, the international community can speak up against it."

Carmen Marsh, based in the United States, shared an incident on privacy that illustrates the importance of placing someone in charge of cybersecurity within every company. The California Earthquake Authority, for example, developed an in-house policy and claims-management system called Epicenter.

At one point, the software development team deployed the upgrade into production, however, some of the access privileges weren't considered. The database included policyholder information from 20 large U.S. insurance companies. Because no one tested the privileges or even thought about privacy, every company could see all the policyholder data of every other company in the database. The issue was fixed, but a big lesson was learnt. The company now has a CISO and an information security manager involved in the software development life cycle (SDLC) to prevent such a mistake from happening again.

These days, there are many private companies that collect an incredible amount of data where there is no supervisory board, governance with controls, or policies on what can be done with the data.

On top of that, there's artificial intelligence and pattern recognition. Swinkels gave an example of how he went to see his dentist and received a Google notification concerning the time he would need to travel there. Whilst Swinkels understands that the dentist is trying to use his data to better serve him, he wonders about the magnitude of Google's data collection tactics given that his appointment was not even in his calendar. Is there, perhaps, another reason why they are monitoring his patterns of movement?

Under regulations such as GDPR, for example, WhatsApp data is not supposed to be shared with Facebook and vice-versa. But who actually verifies this?

"We are talking about private companies with such an enormous amount of data. If it gets into the wrong hands, it can be extremely tricky. And this is where cyber risk comes in," Swinkels noted.

ALWAYS BEHIND THE GAME

Unlike CISOs, black hat attackers are not wed to organisational processes. This means that information security teams are already a step behind.

As CISOs, we need remember that even in white hat hacking environments, when it is discovered that a platform can be breached and the vulnerability reported, the issue is prioritised. The fix will appear in the next patch in a couple of days. Unfortunately, more often than not, the fix gets pushed further down the priority list. The actual time from the detection of a vulnerability to fixing it might take anywhere between four to six weeks.

This unfortunate situation is well known by the black hat hackers. By the time we find the vulnerability, it is already too late. The bad guys have already been sitting in our systems. And whilst the good guys are working tirelessly to fix issues, the bad guys are already working on other exploits.

The question to CISOs and leaders in the security industry is instead of being in a position of constant reaction, how do we lead from the front?

I remain a firm believer that people are our greatest assets. I see lessons in their experiences. As Otto von Bismarck said, “Only a fool learns from his own mistakes. The wise man learns from the mistakes of others.”

Early humans relied on survival instinct. Human nature is ingrained with an independent streak. Early people had to be strong, to find solutions, and solve their own problems. These are the traits still celebrated in modern society.

However, in our industry, to stay ahead of the game we need to collaborate. It was precisely this conviction that led to my journey and a discovery phase, to single out and identify CISOs’ stories to be learned from. This resulted in the book you now hold. “Counsel in the heart of man is like deep water; but a man of understanding will draw it out.” (Proverbs 20:5 KJV).

Have you wondered about the lessons we can learn from how the military stays ahead in the cyberworld? I was certainly curious.

"I was deployed with the Dutch National Police to Iraq after the collapse of Saddam Hussain's regime and dictatorship," Swinkels told me. "I was part of the security reform that rebuilt the Iraqi police and security services."

Swinkels led the transition to intelligence-led policing and was instrumental in many projects and investigations within the counterterrorism, serious organized crime, and border management domains. I asked him about the difficulties he encountered during eight years of building up the Intelligence practice from scratch.

Swinkels thought for a moment then said, "To make a change happen and to innovate, that's hard. We had to rally political support to get us the budget we needed. I was working together with personnel from Germany, Belgium, and Holland. At that time, there was only a single computer at a police station. Everyone was dialling in to the same number just to get information from that one computer."

Swinkels went on to talk about the mindset shift from incident response to the intelligence-led approach to safeguarding society. With the appearance of new technologies, from mobile phones to automatic license plate recognition to CCTV, Swinkels believes analytical tools empower police and intelligence services.

In the next chapter, we'll see how Swinkels helped successfully create change that brought innovation to the intelligence force at a national level, and we will read about his experiences with one of the most successful and well-known American data analytics companies.

YOU DID WHAT?

A great deal can be learned from NotPetya. Tony Vizza of $(ISC)^2$ tells the story of a large manufacturer of building products that was crippled by a variant.

This organisation patched all their systems in response to the original Petya ransomware and updated all its ISMS policies to deal with additional ransomware. They had also updated their Disaster Recovery (DR) and Incident Response plans.

The organisation unfortunately did not disable Server Message Block (SMB1) on servers and workstations. It was also not realized that system administrator accounts had been compromised months earlier, and active testing of DR strategies did not go beyond tabletop exercises.

The ransomware propagated across 70 servers and 1,500 workstations within 15 minutes. Offsite backups were not up to date. Many of the workstations were remote and located in far-to-reach places. The third-party provider contracted to help mitigate the breach was a large national telecom. It had a number of its allocated engineers resign halfway through the clean-up and were unable to fully restore systems. The manufacturer required an additional contractor to assist.

This underscores that when an organization performs a threat remediation activity, all steps should be checked, backups confirmed, and DR strategy tested. Additionally, third parties should never be entirely depended upon to help mitigate a breach.

15

CHAPTER FIFTEEN

Global Perspectives: Staying Ahead (Part II)

DRIVING CHANGE

I was curious to learn Simon Swinkels' biggest takeaway from eight years spent in intelligence-led policing.

His answer was, "I learnt so much from change management, from adopting new technologies and building teams and skill sets that we didn't have before. But do you know what the hardest thing was? The hardest thing to do was change—to convince people to do something different than what they did today."

He gave an example. "I bought a new analysis system for the

national police." It was an innovative solution with a collaborative mechanism enabling different teams to work together across multiple levels. It sounded fabulous to me, however, Swinkels recalled, "I literally faced an uproar from the analyst community."

These analysts had been working with the same tool for the past five years. Some of them had more 50,000 queries pre-built on their machines. The analysts were afraid their of queries would be jettisoned and they would have to change the way they were used to working.

It would have been easier for Swinkels to give in and leave well enough alone. However, he knew that the analysts did not use 80 per cent of the queries. Also, the operational know-hows were stored in individuals' computers. It was a system of knowledge that gave individual analysts status and position, instead of contributed knowledge being passed on for everyone.

This "dependency knowledge" left people reliant on others before they could become effective. Through the tool Swinkels imposed, dependency knowledge became organisational knowledge.

Despite the benefits, change management takes time. Swinkels said it took three to six months to convince the organisation that using the new tool was for the best.

The European model of a military organisation is such that orders are always questioned and decisions analysed. Military hierarchy is based on the principle of tasking empowerment rather than giving orders, which will be covered more in the next segment, Lessons from the Military.

I asked Swinkels to break down how he affected this change. I wanted to learn by understanding his actions and the steps he took during the oppositional blowback.

Swinkels was obliging, "I had to go down to the local

units and explain my decision. Eventually I even gave them a transition period. I said, 'Maybe I made a mistake.' I offered to let them use their old tool in parallel with the new one for six months. If after the trial period the units still feel that the change was a mistake, I would take the blame and they could keep using their current tool."

Swinkels continued, "After two weeks, the chief analytics officer came to me and said, 'You were right.' He went on to become one of the tool's advocates—not because the boss said this was a valuable tool, but it was because it was his expert opinion."

At the end of the day, CISOs must take risks. They should not be afraid to be vulnerable and admit that they might have made the wrong choice. If we want bold outcomes, we need to be bold and dare to make bold choices. We should consult with others, but we might not always have time to wait on other's opinions.

The reason why most start-ups fail is because they don't make choices and unfortunately, if the organisation does, it doesn't stick to it and loses focus. Another reason for failure is a lack of leadership team cohesion. The leaders all have their own dreams and opinions that remain uncommunicated. This results in everyone being pulled in different directions.

START-UP'S SECRET SAUCE

"So Sander, what should a CISO look for in a start-up?" I asked.

"At that time when I joined Palantir as an SEA employee, nobody had heard about us outside the U.S. We were literally a small start-up with no name, no brand," Swinkels replied.

Palantir grew because of trust building and flawless execution. It wasn't just installing a platform and selling a product, rather it was delivering positive outcomes and impact, almost on the spot. In this way, Palantir's marketing buzz went viral and potential clients began to approach them in droves.

Swinkels was extremely proud of the daily excellence of the engineers and the rest of the team executing projects. In his words, "The execution team was able to kick-ass and showed results in short amount of time." To this day, Swinkels still thinks of them fondly. "Working with such brilliant, ambitious, and hardworking individuals was one of the best experiences of my life."

At another time, I had the opportunity to ask Uzair Gadit, one of the most accomplished entrepreneurs in this industry, for his secret to success. "One thing is to establish the right team early on. Many co-founders or entrepreneurs, whilst living in the heat of the moment, don't focus that much on the early days," he noted.

"Secondly, do not shy away from hard work. At times, there is luck at play, but don't get addicted to it. Focus on hard work," Gadit stressed.

"Finally, have your cash generation diversification concern rooted from day one. The search for stability from within regarding present and future cash flows is what primarily led us into many industries, one after the other."

LESSONS FROM THE MILITARY

There are takeaways from the military that we can apply to our industry. Swinkels explained to me that the biggest misperception about military organisations is that people think it's hierarchal and about following orders. In the corporate world, however, people follow a structure that sets out who has what kind of say.

The biggest strength of the military management model is not knowing who can call the shots, but about having a deep trust in peers, subordinates, and those of higher rank to act out the commander's intent.

It is all about ownership or empowerment, not about following orders so much as interpreting what your boss and the next boss up want to achieve. If you don't understand and act on this, you can lose the war.

When basic analytical steps to plan an operation are not followed, there's a high risk of failure. Napoleon and Hitler both went to Russia and each forgot the most simple element each military commander is trained to notice: the weather conditions in which the battles would be fought.

YOU DID WHAT?

I shall end this chapter with a rather light-hearted story Dan Lohrmann shared with me, which I think illustrates a certain point well.

During his many decades of security service, Lohrmann spent 17 years with the U.S. State of Michigan. This included stints as a departmental CIO, CISO, and CTO. In 2009, he was named one of Government Technology's Top 25 Doers, Dreamers, and Drivers. In 2011, Michigan took home five of the ten annual awards bestowed by the National Association of State CIOs.

Lohrmann said that although he was a CTO, he prefers working in security. He told me that security is a positive force that enables other things to happen. That's an important lesson, but one he didn't learn until midway through his career.

Years ago, when Wi-Fi was new, "There were all these

stories about wardriving and people sitting in parking lots and breaking into Home Depot and Lowes and all these places," he explained. He saw nothing but security problems with this new technology. Then his executive team, led by CIO Teri Takai, decided to install Wi-Fi in all conference rooms.

"We had this meeting and I said, 'We can't do it. Not secure. Bad idea. I'm vetoing the project,'" Lohrmann recalled saying. Teri Takai then said, 'I want everybody to leave the room but Dan.' So, they all left. She then said to me, 'Dan, if that's your answer, you can't be the CISO in Michigan.' And I was stunned."

Dan continued, "She said, 'Let me tell you something. I understand all these problems. I understand all these things you're doing, I understand why you're saying it, but I've been to Dow, Ford, Chrysler, and GM, and they all have Wi-Fi in their conference rooms. So you need to figure out what they know that you don't know and then come back and tell me how we're going to do it securely. And we're giving you one week.'"

This triggered a real paradigm shift for Dan, changing his outlook entirely. He kept his job and the State of Michigan's CISO learnt from his private sector partners. They not only installed Wi-Fi, they won several awards for their Wi-Fi two years later. This is not just for being a security enabler, but also for merits that included the process and technology components of the project.

This same man also went on to achieve many remarkable milestones in his career, including advising senior leaders at the White House, U.S. Department of Homeland Security, and other Fortune 500 companies. In less than three years, he led his team to establish a new Michigan Cyber Initiative that included building the Michigan Cyber Range. This was the first non-classified cyber range that has provided training for 50,000 end users, public or private technical staff, and ten million citizens.

Michigan's was also the first state government to build a cyber disruption response strategy. It merged physical and cybersecurity into one office, as well as deployed the federal government's network monitoring system, EINSTEIN 1—the system mandated by the George W. Bush administration to be used by all federal agencies.

Michigan also establish the Cyber Civilian Corps, which is a group of civilians who are also trained cyber experts. The corps members individually volunteer to provide expert assistance to the State in times of emergency.

16

CHAPTER SIXTEEN

Global Perspectives: Staying Ahead (Part III)

EVERYONE DOES IT

There isn't a community or nation that does not seek information on others for the sake of power or influence.

Sander Swinkels said, "I know this for a fact, every country has had an intelligence service since the concept of countries and war between them existed. Let's not be naïve. Just read Sun Tzu's The Art of War. Every country spies on other countries and, for sure, monitors things within its own borders as well."

Edward Snowden's dossier on his country's electronic surveillance practices included information on the NSA's internal surveillance on American citizens, world leaders, and countries such as France, China, and Germany. The United States is not alone in these activities. Every nation state is involved in intelligence gathering.

Swinkels admits he enjoys a good laugh when he reads media fuss about countries that spy or try to influence politics and elections. "You would have to give up on your own societal safeguards if you believed it shouldn't happen. Politicians who deny being involved don't do so only because of political sensitivities, but mostly to avoid appearing stupid or unprofessional if they were to ever get caught."

Swinkels states that he thinks the news frenzy about Russia trying to influence the last U.S. election is a joke. "Honestly, the U.S. has mastered and executed the same skill for decades. Common people should cut through the crap and the news that they read. Russia spies and China does, too. So do the United Kingdom, Australia, Vietnam, Belgium, and even small states like Monaco, the Vatican, and Belize—everyone!"

The real question is how this information is retrieved and for what purpose.

Countries use intelligence, including that which is gleaned from allies, yet they do not trust that intelligence 100 per cent. A government CISO based in SEA said that if he were to use another country's intelligence product, he would be conscious that it would not reflect much of the vendor nation's intelligence. Likewise, he continued, the bias would be equal if he were to use a different country's threat intelligence. He might get a varying picture of external intelligence, as compared to the data the first country would release. As a nation, they have to leverage different intelligence products including open-source data, whilst at the same time safeguarding their nation's information and intelligence. Because of the Internet, things that would have previously remained hidden are now tracked and traced. Further, public awareness has increased exponentially on everything from the possible shady dealings of Donald Trump to the doings of Cambridge Analytica. The good thing about this is that it triggers public debate.

The purpose of GDPR is to give Europeans the authority to learn what is being done with their data and to question why. This is an important first step. If moved in the right direction, without political influence, the public debate could help create a global community of social order.

Swinkels noted, "In Dutch, we have a saying: 'It is like having a block of butter on your head. It will melt anyway.'"

GOING BANANAS

In my current role, I work daily with the red team and white hat hackers. These are the people who plot elaborate schemes to elicit sympathy from people to break into highly secured buildings after hours and may even resort to dumpster diving to retrieve highly classified information. I need to issue a disclaimer here: Our red team only performs these engagements with the proper authorisation and approval from the company. The objective of red team exercises is solely to test current security measures to determine the weakest link and raise awareness as a result.

A company CEO told me a story from years ago—before hacking was commercialised. A particular group of hackers hired a marketing company to do a campaign for Chiquita bananas, One result was a big Chiquita event held in a popular city square in front of a range of major global banks.

Everyone was in party mode. Loud Caribbean music played. Girls in bikinis offered free drinks and gave away bananas and more than two thousand free compact discs.

The real reason this event was organised was so that local bankers would go back to their work desks and insert one of the free discs into their computer. When they did this, a virus was downloaded on their computer systems. As a result, the option to insert CDs was removed from the bankers' computers.

The point of the story is this: We are always a step behind. When an incident happens, only then are policies reviewed and changed.

NOT AGILE ENOUGH

Sander Swinkels advises that the only way to get ahead of the game is to become entrepreneurs ourselves. Think big, start small, and scale fast, he says. Empower the people surrounding us to do things in a different manner.

In an interview with WIRED, Frank Abagnale, who was portrayed by Tom Hanks in the acclaimed movie Catch Me if You Can, commented that what he did more than fifty years ago is actually four thousand times easier today.

Information was hard to come by fifty years ago, he told WIRED. "Technology breeds crime and we are constantly trying to develop technology to stay one step ahead of the person trying to use it negatively."

INTELLIGENCE-LED POLICING

CISOs need to get smart about pre-empting as compared to reacting. CISOs must no longer think only about security measures and monitoring ongoing events. They must think of ways to be smarter than attackers in order to successfully stop more crimes. Intelligence-led predictive policing makes it much harder for an attacker to get in. This alone, can be an effective dissuasion.

CISOs know they don't have the resources and funds required to sufficiently protect their organisations. This is why they must rely on others to obtain intelligence on the latest threats, breaches, and attack trends. It is crucial that threats and intelligence are gleaned from the wider community—for example, by talking to computer emergency response teams from various nations to deduce the country's cyber state, determine the likelihood of attacks, and analyse the patterns, frequencies, and sophistication levels of threats.

I met with Sidney Pearl when he was in Australia teaching proactive cyber threat hunting. Pearl, who is based in the United States, serves as IBM's global leader of i2 Cyber Visual Analytics. For more than two decades, i2 has helped organisations around the world identify threats and stop attacks before they occur. Pearl used to serve with the US Navy in various Special Operations, Intelligence and Communications capacity, and worked closely with the US Secret Service, FBI and Interpol. He presently engages with global CROs, CISOs, and directors of SOC operations to educate them on the integration and application of intelligence in cyber threat hunting.

Whilst security professionals are focused on writing policies and procedures, as well as working diligently to protect their organisations, threat actors are exploiting their vulnerable defences through the use of business models like malware-as-a-service and ransomware-as-a-service. The threat actors are so far ahead of most organisations' traditional defences that "our ability to plug multiple holes become overwhelming when you are simply responding to incidents," Pearl explained to me.

To transform from a reactive to proactive model of operation, it is important, through the use intelligence tradecraft, to ascertain whom the threat actors are and how they're operating. By doing so, answers to the questions of who, what, where, when, why, and how can be learned and the intelligence applied to the cyber kill-chain methodology.

As I have mentioned, I work closely with our own red team across the Asia-Pacific region and the two key things it always endeavours to understand first is what needs to be protected and who does it need to be protected from? Ninety per cent of our red team engagements are focused on the reconnaissance phase including open-source intelligence gathering to social engineering and the actual rogue intrusion. It is by gathering intelligence that the threat actors are able to select their target and plan their mode of attack using the most relevant scenarios.

Smart organisations deploy red teams not to assess the vulnerability of their environments but to determine their weaknesses.

LET THE INTELLIGENCE SPEAK

Currently, the term cyber threat hunting is employed broadly and creates confusion about what it actually is. Pearl defines cyber threat hunting as the human-led, machine-assisted act of proactively and aggressively identifying, intercepting, tracking, investigating, and eliminating cyber adversaries as early as possible in the cyber kill chain.

The goal is to leverage cyber threat hunting to help analysts, typically with intelligence analysis experience, conduct faster cyber incident response and investigation within the SOC and proactive cyber threat hunting for advanced cyber threat hunters.

Most SOC analysts are trained in forensics, however, they do not come from the traditional intelligence background. Hence, there's a huge gap in skills between a SOC engineer or a SOC analyst and an intelligence analyst.

It is the intelligence analysis that most needed. Someone with an intelligence background knows how to gather and access information, focus and transition it to the right outcome, and simulate the attack. They can ask the right questions regarding who, what, where, when, why, and how. These skills will help organisations stop threats more effectively.

How do you communicate the value of cyber threat hunting to CISOs? According to Pearl, "CISOs tell me, they don't want to talk purely about technology. What they want to understand is how to build a proactive cyber threat hunting practice and team. At the moment, fraud is not talking to cyber, and vice versa. With criminal activity acting in the middle, how can they bridge the gap through methodology, tradecraft, skills, intelligence and technology and gain insight into threat actors and threat vectors? They also want to know how to apply the intelligence to the cyber kill chain and shut them down before they attack and cause any problems. The CISOs want to know how to communicate this to the CRO." It is important for organisations to stay ahead of criminal element, instead of always catching up.

Pearl walked me through his process:

1. Do not try to lead with technology.
2. Start with the methodology.
3. Apply technology together with data and intelligence sources. The result will be discovering the right questions to ask.

Most boards of directors now fully acknowledge and agree that cybersecurity is a responsibility that must be addressed, and breaches must be prevented before they occur. The market is coming around to this concept of proactive cyber threat hunting.

There are vendors slowly coming online to provide managed threat-hunting services. Companies do now ask about the use of cyber threat hunting to address the cost of breaches, aid public relations and interaction with the press, and address the lingering effects of time upon the customer base. Organisations who are not should question why they are waiting.

Interest in cyber threat hunting is growing across the Asia-Pacific region, Latin America, Africa, and other areas around the world. Even within the U.S. and European markets, which are perceived as more mature, the danger is that traditional incident response investigation is the final solution. It invokes a false sense of control that many countries are now coming to terms with.

In summary, cyber threat hunting requires a different perspective than traditional incident response and investigation. It is an out-of-the-box, intelligence analyst-driven methodology with a combined data sources and proactive technology approach to stopping attacks before they occur.

It is not possible to stop an attack if a highly skilled hacker puts his or her mind to it. The question: then is this If we know it is a losing game at the get go, instead of investing time in trying to ineffectively pre-empt an attacker, would it make more sense to first invest in a full-fledged purple team?

#Purple Team

So, there's blue teams, red teams, and now purple ones as well?

Yes, in fact, there are.

Blue teams are organisations' defenders and protectors. They ensure security of information systems, identify security flaws, and verify the effectiveness of each security measure.

Red teams originated with the U.S. military in 1984. They were first known as "red cells." Richard Marcinko was the first commanding officer. He was previously also the first commander of the U.S. Naval Special Warfare Development Group, also known as SEAL Team Six.

Marcinko's red cells would target military bases using false IDs, jump fences, and take hostages, including kidnapping high ranking officers. They planted bombs and took over submarine bases. The red cells were extreme in finding weaknesses and videotaped these that tested the effectiveness of U.S. tactics.

In the corporate world, we take it down a couple of notches. First, we must obtain the proper authorisation from a business sponsor or a company executive before planning the reconnaissance. The purpose of these red team engagements is to create awareness and obtain security buy-in at the board and executive level once the members see the extent of exploitable weaknesses.

Every individual in the red team has a unique area of specialisation. Together, they can break into any of the organisation's defences through penetration of the weakest link. They may social engineer their way in by pretending to be maintenance workers or auditors. They have used gadgets such as drones and night vision goggles, and even engaged in dumpster diving to exfiltrate information.

And if you saw what I have through the eyes of my white-hat hackers, you would realise that it's hard to keep up with the countless techniques and attack vectors the bad guys can use as an entry.

CHAPTER SEVENTEEN

17

Global Perspectives: Staying Ahead (Part IV)

THE SILVER LINING

It might seem formidable to play the "staying ahead" game with the bad boys. However, there is a silver lining to it, as seen in the next few short case studies.

The first known reported phishing attack against a bank happened in September 2003. In 2007, the entire nation of Estonia was taken offline during cyberattacks. The latter also occurred in North Korea in 2014. However, public access to the Internet is not allowed in that country so the attack impacted only 1,024 external-facing IP addresses and just four networks.

Since its was attacked, Estonia has become a global cybersecurity superpower, while North Korea has recently been called a hacking titan with a cyber army of an estimated 6,000 hackers. Handpicked by Pyongyang's cyberwarfare agency, Bureau 121, they are trained to plunder international banks, conduct military espionage, and attack foreign nation states' critical infrastructure.

In the commercial sectors, banks are frequently attacked, therefore many have created the top-notch security cultures and cybersecurity teams. The nations, organisations, and industries that tend to be good at cybersecurity are the ones that have no choice.

"A good thing is that awareness is up," Ian Yip assured me. "The average person knows what cybersecurity is. In the past, I had to explain what my job was. Now, C-levels and board members are aware. The World Economic Forum now lists cyberattacks near the top of their global risks, which is useful. Awareness is a good tick."

Yip highlighted that there's a distinction between awareness and education. "Whilst awareness is up, education leaves a lot to be desired. That being said, awareness is the first barrier to getting educated."

As Niccolo Machiavelli, the famous fifteenth century politician and historian, wrote, "There is nothing more difficult to take in hand, more perilous to conduct, or more uncertain in its success, than to take the lead in the introduction of a new order of things."

"I actually like CISOs who are backed in the corner and do not have much budget," claimed Swinkels. After I raised my eyebrows, he continued, "Mind you, this is completely different from the CTO or the IT director, whose budgets have not been too much of an issue at all. With the CISO who has a limited budget, he or she has to be creative in spending wisely. This becomes an amazing opportunity for the start-up community." Such scenarios provide start-ups with great solutions a decent chance of getting into big corporates.

YOU DID WHAT?

(ISC)^2's Tony Vizza tells a story that deserves to be in the You Did What? Hall of Fame.

"When my family and I moved to our current home, it was prior to the completion of renovations. One of the things we didn't have a chance to finish was to install tiling at the front door step, which meant that a gap of about 5-6 centimetres existed between the door and the concrete and left the cavities under the floorboards exposed. It was the perfect sized gap for a rodent to get in.

After about a week, whilst I worked and my partner watched TV, she saw a mouse walk past. She called me down to have a look but by that time it had disappeared. A few evenings later, I was watching TV and saw the mouse. I knew we had a problem." Vizza paused then stressed emphatically, "We keep a clean house and have never had rodent issues. Not once!"

He continued, "Our first approach was to try to humanely capture the rodent and let it go far away from the home, as we did not want to kill it. The mouse toyed with the humane trap. It would pinch the food inside the trap and not allow itself to get caught. When I started to find mouse droppings around the home, I decided we needed to up the ante and went for a lethal mousetrap. I deployed the trap before going to bed and, lo and behold, caught the mouse. I was actually a bit saddened that I had killed it, but I disposed of it and thought 'That's sorted. No more mouse.'"

A few days later, Vizza was again watching TV when another mouse scurried by. "Trap deployed again, caught another mouse, disposed of mouse. Just to be safe, I reset the trap and left it there. I then went to work. An hour later, I got a phone call from my partner who was cranky that I had not disposed of the mouse that had been caught. I said, 'What do you mean?' and she said, 'There's a mouse in the trap!' Mouse number three. In the end, we caught four mice."

What they discovered throughout this process was:

1. The mice entered the house through an untiled gap at the front door and others under the floorboards.
2. The mice had feasted on a pack of oats that sat forgotten at the back of a cupboard.
3. The mice had also found a box of Easter Egg chocolates that had been hidden from the kids when they had a nice party.

To remediate, Tony had to:

1. Close up all gaps in the kitchen.
2. Remove all gaps where there was cabling that went under the floorboards.
3. Deploy traps.
4. Dispose of the contents of all of the food in the cupboards because of the risk of contamination.
5. Literally wash every pot, pan, and kitchen item.
6. Put all food in containers before placing them in cupboards.
7. Have an external pest controller ensure all mice were gone.
8. Clean the house from top to bottom.
9. Install tiling at the front door to close the gap that was

used by the mice.

Vizza said he told this story at an insurance event a few years back. The CEO of the multi-billion-dollar company came up to him and said, "Twenty-eight."

He had no idea what he meant until this CEO explained that he'd had a similar situation at his home that resulted in 28 mice exploiting a gap in the plumbing.

What is the moral of the story?

1. Hackers, like mice, will exploit a vulnerability and weakness.
2. Hackers, like mice, will only be noticed if they accidently make their presence known.
3. Hackers, like mice, will sit on your infrastructure and wait for the opportune time to cause damage.
4. Hackers, like mice, will consume your resources.
5. You never know when you have fully rid your network of hackers in the same way you don't know if you have rid yourself of mice.
6. To this day, Vizza has no idea if the mice actually removed anything from his home, and it's the same with hackers.
7. The cost associated with having to throw away a pantry full of food was substantially higher than the cost of sealing the gaps.
8. Hackers, like mice, will make you deploy defence in depth after you have experienced the damage caused by them.
9. Deploying defence in depth took substantially longer to implement than having filled the gaps during the renovation.

BONUS CHAPTER

Selling to the CISO

CISOs want to understand how to speak business language to better support organisational goals. The wider world wants to get into the minds of CISOs to better support their cybersecurity goals.

For some CISOs, it is tempting to buy into new ideas, and it seems a natural solution to purchase new firewalls or intrusion prevention systems. However, CISOs are generally not swayed by products or the latest systems. What's most relevant to them are solutions placed within a bigger, long-term picture.

However, there are some essential considerations. CISOs

first focus is in building the basics before moving on to other layers. In the bigger scheme of things, sometimes they don't need to work on the system. It might be more important to heighten the awareness of employees, which would result in raising the resilience profile of the organisation. In such cases, the purchase of a new product would be the last of their priorities.

In choosing security partners, CISOs are mindful that small companies can be competitive in pricing, but the risk is higher of not fulfilling their mission is greater. That being said, CISOs look for the best capabilities balanced with the most reasonable prices.

CISOs appreciate a strong partner. Israel Baron shared how he selected a defence company to build his SOC. He liked the fact that they provided excellent demonstrations that offered clear instructions. The result was an effective transfer of knowledge throughout the whole company.

I grilled a number of CISOs on what they look for when choosing a vendor partner. They told me they look at several factors, including whether the proffered solution integrated with other products already in use, whether it was part of a consolidated system that allowed management from the same dashboard as another, whether the staff would need additional training, and would additional staffers be required to run it? Is the organization mature enough for this solution? Will it require fine-tuning?

A good CISO, I was told, knows he or she will never tick all the boxes, but tries to hit as many as possible. They also know that some tick boxes much more important than others.

There are vendors who are focused on their offerings. Anything more than that is beyond their capability. There are also vendors who will "admit that they do not have the solution right now," one CSO said. "However, they fully agree with me that their vision is aligned with mine, and they are also working in the same direction. They end up working together with me, taking the journey, as we come up with a solution." This is true partnership, when both sides help the other to become successful.

A CISO from Atlanta, Georgia, in the United States, affirmed he has had true partnerships vendors. He said he values the vendor who shares the same goals and supports him with their resources.

Generally, CISOs usually do more listening than talking, but they do not want vendors to tell them what they need. In fact, one CISO intentionally gives his vendors a pain point that they will not be able to resolve with today's technology. He does this to explore how the vendor proposes to be part of his ecosystem, move along with him, and enable him.

A CxO observed that when a vendor says to a CISO, "You need to buy our products," it demonstrates no attempt whatsoever to understand the CISO's priorities, challenges, pain points, or plans for the future.

Tony Vizza of $(ISC)^2$ sums it up, "In a nutshell, CISOs are looking for customer service. When done properly, this means more than the vendor picking up the phone or answering emails quickly. Customer service means honestly dealing with a CISO, and being transparent about deficiencies, delays, and shortcomings."

Vizzo pointed out that CISOs seek consistency. For example, many vendors "churn" staff regularly through restructures or account management reallocation. Meanwhilst, CISOs rely on the vendor partners they have built relationships with.

Dan Lohrmann has sat in hundreds of meetings with vendors eager to supply solutions to the government sector. However, almost every private sector company that sells to or supports government enterprises has been frustrated with the way business is conducted in that sphere. On the other hand, it is also true that frustrations abound on the public sector side of the fence. Dan has heard many CxOs share stories about vendors who "just don't seem to get it."

Next, he and other CISOs offer advice to vendors.

READ MY MIND

Most CISOs cultivate strong networks of industry professionals and companies they know and trust. Lohrmann explained, "If you are not on that list, or even if your company is on the list but you don't know the specific person or team, don't act as if you are life-long pals."

He offered some practical advice to vendors trying to read the minds of government security leaders: Be genuine and don't misrepresent mutual contacts, friends, or experiences. Building trust takes time and it likely won't happen in the first few meetings.

Personally, I've seen vendors walk into meetings that took months to arrange, then ask, "What are your priorities? What's on your mind?" They might think they are being good listeners, but I wonder if they are really listening or already have a solution in mind.

Lohrmann agreed that it is rare to see a vendor with a balanced approach. He said they are usually "all listen or all talk," which is then seen as a waste of time by CxOs. His advice to them is this:

- Do your homework on government needs and requirements in advance.
- Read the potential client's strategic plan.
- Offer a meeting agenda in advance and ask whether the agenda is acceptable beforehand.
- Be clear on the specific topic and solutions on offer at the meeting.
- Understand and respect the busy schedules of government teams.
- After your presentation, ask for input such as, "What do you think? Does this meet your identified needs? Where can we improve?"

At the end of the day, CxOs are looking for a trusted partner. They are pondering whether your company is in for the long haul or a quick buck? Can you be trusted? Do you have a good reputation? Can you deliver? Will you be around in two years? Do you have case studies proving what you have done for previous clients?

These are just some of the questions on the CISOs' minds. Most usually have a mental map of their requirements and the support and capability they are looking for. Some might have metrics to determine whether the solution is a strategic fit for their organisation.

As an executive advisor who has worked directly with countless CISOs, including running many CxO panel discussions and talks, my key advice for building strong lasting relationships and trust is to keep it authentic. Be genuinely interested and focused on meeting the CISO's business needs, and always maintain the integrity of your conduct and conversations.

THE BIG NO-NOS

When it comes to using negative media reports for the purpose of gaining credibility or to push the security agenda, here's the thoughts of an executive after a vendor presentation: "It was so painful. So painful. Everybody knows how painful it is. Trust me. Please, to every vendor partner and product developer out there, stop it."

Bad news doesn't work. "Don't use gimmicks or scare tactics when presenting findings. Focus instead on risks and measures to mitigate vulnerabilities." RenaissanceRe's Kelly Isikoff opined.

YOU DID WHAT?

This is the last You Did What? and what better way to end this book!

I founded Cyber Risk Meetups in 2017. Since then, the group has blossomed to approximately 1,800 members and now holds events in Singapore and across Australia.

Renowned for running CxOs panel sessions and uncensored talks where attendees hear straight from the mouth of executives, Cyber Risk Meetups became oversubscribed for any event we ran, easily hitting more than 200 registrations per iteration.

It was during one of these Meetups that I was introduced to Steve Simpson, the former chair of the Australian Information Security Association's Perth Chapter. Simpson served in the British Army as an information systems manager and once managed three teams of IT and communications specialists.

Simpson discussed being deployed to Afghanistan in

2002. He and his compatriots on the IT team and associated systems lived in tents in rough conditions. Steve remembered a fine day when he went to the IT area first thing in the morning to make sure everything was ok. One of the night shift personnel appeared to tell Simpson that he'd missed out on all the excitement. Every device in the HQ had a virus and it took the whole team all night to resolve. He was happy to report that there wasn't any downtime at all.

When Steve investigated, it turned out that this guy had received an email from his wife that contained instructions to check his systems for a file. The icon of a teddy bear was an indication of the bugbear virus.

For those who remembered the jdbgmgr.exe virus hoax, it involved email spam advising computer users to delete the file as it was a computer virus. In actual fact, this was a normal debugging file in Microsoft windows. Unfortunately, Simpson's staffer, well trained as he was, believed the hoax email sent by his wife and got his team to spend the entire night removing the Microsoft file on every single system.

Fortunately, the hoax pointed at a file few people used. If it had been pointed at part of the boot sequence of the operating system, then they would not have been able to reboot the system and the outcome would have been very different!

Lesson learnt that day: Not everything you received is true, even from a trusted source . You need to verify it.

With that, we have come to the end of this book. We've come face to face with more than thirty CxOs from around the world. They have graciously supplied life lessons that are summarised herein. As we equip our current and next generation of leaders, let us never forget the value of continuous learning and collaboration. It is whilst we embrace our own failures and the mistakes made by others, that we become stronger as a whole.

Onwards and upwards!

Shamane Tan regularly runs CxO panel discussions and gives international talks to spread awareness on cyber risk topics. Reach out to her at ***shamane@cyberriskmeetup.com*** *for engagement queries.*